LIFTOFF

ALSO BY MICHAEL COLLINS

CARRYING THE FIRE
An Astronaut's Journeys

LIFTOFF

THE STORY OF AMERICA'S
ADVENTURE
IN SPACE

Michael Collins

ILLUSTRATED BY
JAMES DEAN

GROVE PRESS
New York

Published by Grove Press
a division of Wheatland Corporation
920 Broadway
New York, N.Y. 10010

LIBRARY OF CONGRESS CATALOGING-IN-PUBLICATION DATA
Collins, Michael, 1930-
Liftoff: the story of America's adventure in space / Michael Collins: illustrations by
James Dean.—1st ed.
p. cm
Includes index.
ISBN 0-8021-1011-8
1. Astronautics—United States—History.   I.Title.
TL789.8.U5C59 1988   88-1706
629.4′0973—dcl9   CIP

Designed by Richard Grant

Manufactured in the United States of America

First Edition 1988

10   9   8   7   6   5   4   3   2   1

For Ted, Charlie, Roger, C.C., and Donn—
Our group of Fourteen is now Nine.

AUTHOR'S NOTE

This book began in the mind of Howard Golden, Educational Publications Director at NASA Headquarters in Washington. Howard wanted people to understand how spacecraft and space equipment worked, and know who had designed and built them. The American public had heard about astronauts, but that was all. Howard wanted a book that explained the rest, from the early days of Project Mercury to the lunar landings, and asked me to write it. I agreed, with the proviso that I would do it my own way, without anyone looking over my shoulder. *Liftoff* is the result.

In writing this book I found the most pleasure in renewing old friendships. The machines themselves were worth revisiting (what *do* they do with all that carbon dioxide you exhale?), but beyond that were the people. What made them so special, a group brought together at a pivotal time, one that performed beyond reasonable expectations? Why were they so successful?

Successful that is, until the Challenger explosion. How did the space program fall so far, so fast? That too needs discussion, in terms of NASA's present unhappy condition and—more important—what reasonable possibilities exist for the future. I hope this book answers some of those questions.

In writing *Liftoff* I have become indebted to several dozen people—those who have shared their memories with me, and those who have commented on what I have written. The responsibility for the final product is mine alone, but they have strengthened it immeasurably.

Lee Saegesser, NASA archivist, was particularly helpful in digging into his voluminous files, as was Air Force historian Richard Hallion in sharing his prodigious resources.

Bob Gilruth, Max Faget, and Caldwell Johnson helped me with the early days of NASA, as did John Yardley and Mike Weeks for industry's viewpoint. As NASA progressed, Chuck Mathews, Joe Gavin, Joe Shea, and John Healey picked up the story, and Tom Paine provided an administrator's overall perspective. Julian Scheer provided not only recollections and advice, but office space as well.

Flight experiences from Mercury, Gemini, Apollo, Skylab, Apollo-Soyuz and the Shuttle are included thanks to astronauts John Glenn, Pete Conrad, Jim Lovell, Al Bean, Joe Kerwin, Bill Pogue, Jerry Carr, Vance Brand, and Joe Allen. And beyond talking about it, John Young, Neil Armstrong, and Buzz Aldrin deserve special thanks for putting up with me in flight.

Helpful friends who have read parts of the manuscript include Fred Asselin,

Silvio Bedini, Howard Paine, Ben Schemmer, Ken Thornton, Paul Wagner, and Jack Whitelaw.

Ian and Betty Ballantine, like Howard Golden, have believed in the book from the beginning, and have been wonderful, steady mentors. Richard Ballantine, a most supportive editor, deserves extra thanks as do all those involved at Grove Press and Weidenfeld and Nicolson, above all Ann Getty and Fred Jordan.

Two special friends, my agent Maria Downs and typist Terry Pietroski, took care of things that are beyond my ken. My wife Patricia cheerfully ignored my moody silences during the two years of the book's preparation, and I really appreciate that.

And a final thanks to my collaborator Jim Dean, who is not only one of the finest artists I know, but someone who understands NASA and spaceflight very well. It was a real pleasure working with him.

To any people I have overlooked, my apologies.

Michael Collins
Washington, D.C.
January, 1988

ILLUSTRATOR'S NOTE

I am grateful to many individuals and organizations in government and industry, who helped me with the pictorial research necessary to create the illustrations for this book. A number of drawings would have been extremely difficult, if not impossible, to do without the help of knowledgeable professionals at the National Aeronautics & Space Administration, particularly Marie Jones and Lee Saegesser at NASA Headquarters, and Vance Brand and Mike Gentry at the Johnson Space Center. It was essential for me to see, and sketch, some of the historic spacecraft on display at the National Air & Space Museum, as well as refer to the Museum's library a number of times. For this assistance I'm deeply appreciative. McDonnell Douglas Astronautics Company graciously gave permission to use reference material from the early manned spaceflight programs. My thanks go to Mike Collins, who looked carefully at each illustration and sometimes said, "Jim, that's not *quite* the way it was," and then gave me his expert suggestions.

And finally a special thank you is due to Paul Calle, a superb artist, whom I've known for nearly twenty-five years. By his work he has taught me a great deal about drawing.

James Dean
Washington, D.C.
January, 1988

CONTENTS

ACRONYMS AND KEY TERMS

BIGs	Biological isolation garments.
CM	Command Module. Mother ship.
CSM	Command and Service Module. Command Module and attached Service Module. Crew capacity: three.
ECS	Environmental control system regulating atmosphere inside spacecraft.
EOR	Earth Orbit Rendezvous Mode. Two independently launched spacecraft joined together while in earth orbit.
EVA	Extravehicular activity in space, or "space walk."
HOUSTON	Manned Spacecraft Center.
HUNTSVILLE	Marshall Space Flight Center.
LM	Lunar Module. Two-stage moon landing craft. Lower stage used for descent and then as launch pad for the upper, ascent stage. Crew capacity: two.
LOR	Lunar Orbit Rendezvous. Voyage to moon where spacecraft is a single unit until in moon orbit, and then separates into a mother ship (Command Module) and a landing craft (Lunar Module). Astronauts use the landing craft to descend to the moon, and return to the mother ship. When the mother ship departs for Earth the landing craft stays behind, parked in moon orbit.
LRL	Lunar Receiving Laboratory.
NACA	National Advisory Committee for Aeronautics.
NASA	National Aeronautics and Space Administration, or Space Agency.
SM	Service Module. Storehouse attached to Command Module.
SPS	Service propulsion system. Command and Service Module rocket engine.
TEI	Trans-earth injection. Start of voyage from moon orbit to earth.
TLI	Translunar injection. Start of voyage from earth orbit to moon.

1

Apollo 11

A space flight begins when the technician snaps your helmet down into your neck ring and locks it in place. From that moment on, no outside air will be breathed, only bottled oxygen; no human voice heard, unless electronically piped in through the barrier of the pressure suit. The world can still be seen, but that is all—not smelled, or heard, or felt, or tasted. Now Neil and Buzz and I are ''suited up,'' ready to begin our voyage. We have gotten up early this morning, shortly after 4 A.M., to prepare for an 8:32 A.M. launch. We have undergone cursory physical exams and enjoyed the customary launch day breakfast of steak and eggs. We have said goodbye to our families and friends and we are impatient to start. Carrying our portable oxygen containers like attache cases, we walk down long corridors and emerge into the piercing Florida sunlight. Along the way coworkers wave and cheer. Although we can't hear what they are saying, their enthusiasm is contagious and we pick up the pace a little, waving back, and then enter a small van for the 8-mile ride out to Launch Pad 39A. Even at 7:00 A.M. the road is clogged with tourists, their cars bumper to bumper. But we have a police escort and our own lane, over on the shoulder, and within minutes we stop at the base of the launch tower.

The pad is different today. Usually it is a beehive of activity, but now the workmen have abandoned the place, for the tanks have been filled with deadly hydrogen and oxygen and no one touches a loaded Saturn V except the crew and a few people to help us on board. The monster rocket steams daintily in the morning sunshine as we enter a small elevator and make our way up to the 320-foot level, the entrance to our Command Module Columbia. We pass over a narrow walkway and I stop there for a moment, compelled by some inner urge to savor the view, to consider the moment. I feel more like an explorer than an astronaut. If I cover my right eye I can see only the unsullied beach, the Florida of Ponce de Leon, and beyond it our primordial mother, the sea. If I look to my right I see technology and the United States of America and the most complicated pile of machinery ever assembled. I have to confess that something inside me pulls toward the beach, but I have made my commitment to the machine and now it is time to get started. There will be time for beaches after the moon, I hope.

Our seating arrangements are peculiar: we ride in whatever couch our specialized training dictates. I ride on the right for launch, on the left for entry. When by myself in the Command Module, I roam, to the lower equipment bay for navigation, to the right to check the electrical system, to the center for computer work, back to the left for docking. The Command Module is my baby, but Neil is the Commander, and as we await launch he sits in the decision-maker's spot, on the left where the abort handle is located. One twist, and the solid rocket escape tower will blast us clear of the trouble below. My backup, Fred Haise, has preceded us into Columbia and has spent a couple of hours running through a 417-step checklist, making sure that every switch has been set in its proper position. Now we don't have a lot to do except wait, the hardest part of all. I wish they would do away with the formal countdown and just have a pleasant voice whisper in our ear, ''Now!''

At ignition the Saturn V gives us a little surprise. Instead of the hideous din one would expect, it is quiet enough inside to hear the radio, with our volume knobs set up high. But the Motion! Instead of the stately ascent I have watched from a safe distance there are quick little motions as we leave the ground. I feel our engines swiveling left and right, keeping us posed in delicate balance despite crosswinds and sloshing fuel tanks. It is like a nervous novice driving a wide car down a narrow alley and jerking the wheel back and forth spasmodically.

But in general, compared to some other rockets, the Saturn V is a gentle giant. Granted, the first stage puts out an incredible 7.5 million pounds of thrust, but the whole thing weighs 6 million pounds. Therefore at liftoff, dividing 7.5 by 6, the acceleration we feel is only 1.25G—1¼ times our normal weight. As we ascend, and the fuel tanks empty, the thrust remains constant while the weight of the Saturn V decreases. Following the same formula ($F = Ma$), the acceleration increases gradually and at first stage burn out, 2½ minutes into the flight, we weigh slightly over 4 times our normal weight. No problem, as long as we don't have to reach any awkwardly located switches.

When the first stage separates at an altitude of 45 miles, for an instant we are jerked forward in our straps and then the five J-2 engines of the second stage take over and again we are pushed gently back in our couches. This is the stage of which we have become wary, the stage of brittle aluminum and faulty welds, but today it is perfection. High above atmospheric disturbances now, our climb is smooth as glass, as quiet and serene as a rocket ride can be. There is a protective cover over the Command Module for the first 3 minutes. When we reach an altitude of 60 miles, too high to use the launch escape tower anymore, it is jettisoned along with the cover, and for the first time we can see out our windows. However, we are pointed straight up and there is not much to see, except a small patch of blue sky that gradually darkens to the jet black of space.

When we reach 110 miles, the single engine of the third stage ignites and drives us downrange, increasing our speed to the 17,500 mph we need to achieve orbit. Compared to the second stage, the motion of the third stage is rough, buzzing and rattling, and I'm relieved when the engine shuts down. The ascent has taken slightly less than 12 minutes.

Now we have an orbit and a half in which to make sure all our equipment is operating properly before we re-ignite the third stage engine and commit ourselves to leaving the earth's gravitational field. I am also weightless again for the first time in 3 years and want to enjoy it, but there will be time for that later.

The Saturn has placed us with our heads toward the earth and our lower equipment bay pointed toward the sky, so that I can check out our sextant and other navigational gear. That means we are ''upside down'' in the terrestrial use of the term, but in weightlessness it seems quite natural to be looking ''up'' at the earth. The view is spectacular, with storm clouds over the Pacific, but there is no time to savor the show because I have a lot of work to do in a short time. I take off my helmet and gloves, collapse the lower half of my couch, and ease over it into the lower equipment bay. Despite my impatience, I move slowly because we have been warned about space motion sickness, ever since Frank Borman became ill on the way to the moon, creating quite a commotion in Mission Control.

The theory is that none of us got sick inside a Mercury or a Gemini because the tiny cockpits were so confining, but that the extra room inside the Command Module may trigger a reaction. We have been told to move carefully and not wiggle our heads left or right any more than we have to. (In weightlessness, the fluid in the inner ear can slosh to and fro freely.) I slide around cautiously, as if I were wearing a sleeping poisonous snake as a collar. I think about my stomach. Is it all right? It seems to be, but just asking the question is the first step toward planting a doubt in my susceptible brain.

Fortunately my body seems as healthy as our equipment. After visiting the far corners of the machine, as proprietor of the Command Module, I declare it ready for the next step. Neil and Buzz agree. Before the appointed time for re-ignition I am back in my couch, with helmet and gloves back on. This last touch is in case the Saturn explodes and somehow dumps our cabin pressure. In that event we would, in theory, be protected inside our sealed pressure suits. However, any explosion massive enough to crack our hull would also result in multiple equipment failures, and we would never get back in one piece. Still, a rule is a rule, so we sit there, helmet and gloves on, ready to be propelled to another planet.

Mission Control gives us their blessing, in the esoteric patois of spaceflight: "Apollo 11, this is Houston. You are go for TLI." TLI stands for translunar injection, increasing our speed to nearly 25,000 mph—escape velocity!

At ignition we are over the middle of the Pacific Ocean. Specially equipped aircraft below us record a stream of information telemetered down to them, and relay it back to Houston. They are bottle-nosed KC-135s, converted Air Force tankers carrying an array of electronic gear. In less than 6 minutes the burn is over, the Saturn V has done its job, and we are on our way. It is less than 3 hours from liftoff, and I know that the highway leading south from Cape Canaveral must still be jammed. Imagine that! We have gone around the world almost twice and people who came to watch the launch are still bumper to bumper struggling back to their motels.

We are climbing like a dingbat and passing through 1,200 miles now. Although we are traveling 6 miles per second, far faster than a rifle bullet, it is hard to tell so by looking out the window. Unlike the roller coaster ride of earth orbit, we are entering a slow-motion domain where time and distance seem to have more meaning than speed. To get a sensation of traveling fast, you must see something whiz by: the telephone poles along the highway, another airplane crossing your path. If there is a blur, so much the better. In space, objects are too far from each other to blur or whiz, except during a rendezvous or a landing and in those cases the approach is made slowly, very slowly. But if I can't sense speed out my window, I can certainly gauge distance, as the earth gets smaller and smaller. Finally the whole disk can be seen. For the first time I think I know what "outward bound" means.

We cannot watch the shrinking earth for long, however, because of the heat coming from the sun. We must turn broadside to it, and begin a slow rotation, so that the solar radiation is absorbed evenly by all sides of our spacecraft. We are like a chicken on a barbecue spit. If we stop in one position for too long, all kinds of bad things happen: pipes freeze on one side while tank pressures increase on the other.

As our spacecraft rotates on its axis, so does the earth. We try to keep our antenna pointed at it, to ensure communications with Mission Control. Of course, Houston isn't always in view, but our voices are relayed through the tracking stations in Spain, Australia, or California. Our next task is a pleasant one, to get out of our bulky spacesuits. We help each other unzip and remove them, thrashing around like three albino whales inside a small tank, banging into the instrument panel despite our best efforts to move slowly. In weightlessness Newton's law of action and reaction reigns supreme. Every time we push against the spacecraft our bodies tend to carom off in some unwanted direction and we have to muscle them back into place. Finally, one by one, we get the suits off and folded neatly inside storage bags under the center couch. Now dressed in white two-piece nylon jumpsuits, our bodies seem much smaller and the inside of Columbia a lot more spacious.

It is a quiet interval and we get a chance to examine our surroundings, this strange region called cislunar space. Is it daylight? Yes, the sun is definitely shining on us. Is it dark? Yes, if we shield our eyes from the sun, the sky is flat black except for faint pinpoints of starlight. With no sunrise or sunset to cue us, we simply ignore the outside world and rely on our wristwatches, set at Houston time. When it becomes bedtime in Houston, we will retire. Our internal clocks, following the circadian rhythm of our bodies, will keep us tied to Houston's work-rest cycle, as if we had never left home.

On the other hand, we can't change the distance between earth and moon to suit our bodies, nor our speed, so we will arrive in the middle of the night, Houston time, and will just have to ignore our drowsiness and press on with our work. In the meantime, we sleep comfortably in light nylon bags attached beneath our couches. It is a strange but pleasant sensation to doze off with no pressure points anywhere on the body, suspended by a cobweb's light touch— just floating and falling all the way to the moon.

During the next 3 days Neil and Buzz have a last opportunity to do some home-work, reviewing their checklists and rehearsing their landing procedures. Housekeeping aboard Columbia is my responsibility, and I have a sizable list of chores: fuel cells to be purged, batteries charged, waste water dumped, carbon dioxide canisters changed, food prepared, drinking water chlorinated, and so on. Our speed decreases steadily: the sun is pulling us, the moon is pulling us, but the earth's tug will remain dominant until we are within 40,000 miles of the moon. At that point we will have slowed from 25,000 mph to a mere 2,000 mph, and then we start to speed up again as the moon's influence becomes dominant.

As we approach our destination we stop our barbecue motion and swing around to study the moon for the first time in 3 days. The moon I have known all my life, that two-dimensional, small, yellow disk in the sky, has gone away somewhere, to be replaced by the most awesome sphere I have ever seen. To begin with, it is huge, completely filling our windows. Second, it is three-dimensional. Its belly bulges out toward us in such a pronounced fashion that I feel I can almost reach out and touch it. The sun is behind it and its light cascades around the moon's rim.

The moon's face is divided into two regions, one nearly black and the other basking in a whitish light reflected from the surface of the earth. This earthshine, as it's called, is considerably brighter than moonshine on earth. The reddish-yellow of the sun's corona, the blanched white of earthshine, and the pure black of the star-studded sky all combine to cast an eerie, bluish glow over the moon. This cool, magnificent sphere hangs there ominously, a formidable presence without sound or motion, and issues us no invitation to invade its domain. Neil sums it up: "It's a view worth the price of the trip."

We must slow down now to enter lunar orbit or we will sail on by the moon in a gigantic arc and return to the vicinity of the earth. We need to reduce our speed by 2,000 mph, from 5,000 down to 3,000, and will do this by burning our service propulsion system (SPS) engine for 6 minutes. We are extra careful, paying painful attention to each entry in our checklist. We have a lot of help, from our computer and from Mission Control, but if just one digit slips in our computer, and it is the worst possible digit, we could turn around backward and blast ourselves into an orbit headed for the sun. When the moment finally arrives the big engine ignites and pushes us firmly back into our couches. Heavy as we are with a Lunar Module full of fuel attached to Columbia's nose, the acceleration is only $\frac{1}{5}$ G but it feels good after 3 days of weightlessness.

Once in orbit around the moon, at an altitude of 60 miles, I have a chance to marvel again at the precision of our path. We have missed the leading edge of the moon by only 300 miles at a distance of nearly a quarter of a million. Like skeet shooters, our guidance experts aimed us at a point in the sky nearly 40° of arc ahead of the moon's position at launch, and the three of us have spent the past 3 days racing toward our rendezvous.

Now we have a chance to examine the back side of the moon, the part never visible from earth, and it is quite different in that there are no flat maria, or seas. It is all rocky highlands, densely pockmarked by 4.6 billion years of

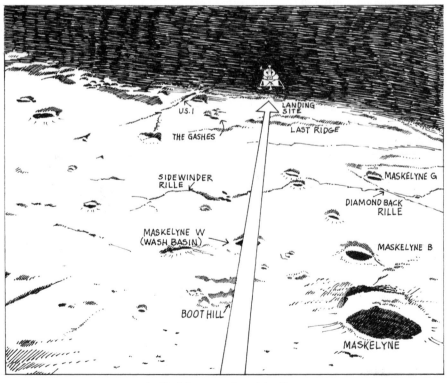

Apollo 11 approach and landing

meteoroid bombardment. We also examine our landing site closely, and Neil and Buzz call out their memorized checkpoints leading to it: Mount Marilyn, Boot Hill, Duke Island, Diamondback, Sidewinder. It is just past dawn in the Sea of Tranquillity and the sun's rays are intersecting the landing site at a very shallow angle. Under these conditions the craters on the surface cast long, jagged shadows, and to me the entire region looks distinctly forbidding. I don't see any place smooth enough to park a baby buggy, much less a Lunar Module.

We have been asked to settle a controversy. The Apollo 8 crew reported that the moon was gray, while the Apollo 10 crew claimed it was brown. Who's right? We decide both: it depends on the sun's angle. Near dawn or dusk we vote with Apollo 8: the surface appears gray, shifting from charcoal to a lighter tone. Near local noontime, on the other hand, with the sun shining directly down on the surface, there is definitely a rosy, tan hue.

Despite the fact that I have spent years studying photographs from Ranger, Lunar Orbiter, and Surveyor, as well as from Apollo 8 and 10, it is nevertheless a shock to actually see the moon at firsthand. The first thing that springs to mind is the vivid contrast between the earth and the moon. One has to see the second planet up close to truly appreciate the first. I'm sure that to a geologist the moon is a fascinating place, but this monotonous rock pile, this withered, sun-seared peach pit out my window offers absolutely no competition to the gem it orbits. Ah, the earth, with its verdant valleys, its misty waterfalls . . . I'd just like to get our job done and get out of here.

Neil and Buzz are preparing to leave in Eagle, donning their pressure suits once more. This time they begin by putting on liquid-cooled underwear, strange mesh garments resembling long johns into which have been woven hundreds of small, transparent plastic tubes. On the moon, cold water from their backpacks can be piped through these tubes to cool them much more efficiently than blowing oxygen over their bodies, the only source of cooling during our Gemini space walks. Back in Columbia I won't be generating enough heat to require liquid cooling.

As soon as Eagle separates from Columbia Neil and Buzz park outside my window and do a little pirouette, turning in front of me so that I can check Eagle's external condition and make sure that all four landing gear are properly down and locked. Eagle doesn't look like any eagle I have ever seen. It's upside down, as my eyes see it, and its spindly legs are jutting up above a lumpy body that has neither symmetry nor grace. Parts of it seem stuck on at the wrong angles. It's the weirdest looking contraption I have ever seen in the sky. But Neil and Buzz seemed pleased with it, and I lie a little in my report to them: "I think you've got a fine looking flying machine there, Eagle, despite the fact you're upside down."

As they separate from me I keep them in sight as long as I can by tracking them with my sextant. Part of this is sentimentality and part practicality, for if they have to be rescued before landing it would help if I could see them.

I strain at the eyepiece until finally at 115 miles they disappear among the craters below. Soon I listen with great concern as they report alarms coming from their computer. A young controller named Steve Bales sits at his console in Houston and fires back reassuring answers as each alarm erupts. We find out later that the computer was overloaded with information flowing to it from both the rendezous and the landing radar sets, but at the time all Neil and Buzz could do was to trust Steve, and they did.

The area picked by the computer for landing does not suit Armstrong. It's littered with boulders the size of Volkswagens, and he decides to keep going and flies over a rocky crater onto a smoother plain. By this time he has only 30 seconds of landing fuel remaining, so he sets Eagle down. Whew! His task was complicated by flying dust. It did not obscure his forward view, and he could accurately judge his descent rate, but as the dust shot out radially from the descent engine, it hindered his ability to tell whether he was moving fore and aft, or slipping to the left or right. He landed in a slight left drift, but gently—a "very smooth touchdown" according to Buzz, who should know. "Houston, Tranquillity Base here. The Eagle has landed," Neil reports.

Our flight plan calls for a sleep period immediately after landing, but the three of us have talked about this before the flight and our opinion is that with all that adrenaline pumping, it would be best to keep going out onto the surface, and sleep later. Houston now agrees, and after a leisurely preparation period, taking nearly 6 hours to ready their equipment, Neil and Buzz open the hatch and Neil backs down the ladder. "That's one small step for man, one giant leap for mankind," he calls as he alights, forgetting to put an "a" before "man."

The lunar surface reminds Neil of the high desert, and Buzz is struck by the "magnificent desolation." One of their first tasks is to unveil a plaque on the leg of the Lunar Module, showing the earth's two hemispheres, with the signature of President Nixon plus the three of ours and the statement: "Here men from the planet Earth first set foot upon the moon, July 1969 A.D. We came in peace for all mankind." They also set up an American flag with some difficulty, unable to poke a hole more than 6 inches deep in the dense, rocky soil. President Nixon joins them in what he calls "the most historic telephone call ever made" and says that "For one priceless moment in the whole history of man, all the people on this Earth are truly one. . . . "

Neil and Buzz practice moving about the surface, even hopping like kangaroos, and then go about their main task of collecting interesting-looking rocks. They also deploy scientific equipment, such as a laser reflector, stereoscopic camera, solar array, and seismic package, for some simple experiments. They sample the subsurface layers by driving a core tube into the ground. Their allotted 2½ hours goes swiftly and then they clamber back into the Lunar Module, shut the door, and repressurize the cabin. Neil reports that "the sights were simply magnificent, beyond any visual experience I had ever been exposed to." Houston has a million questions, and by the time Neil and Buzz have answered them and stowed their equipment, their work day has stretched to 24 hours. Tired as they are, they still sleep only fitfully, cramped and cold in a machine designed for purposes other than relaxation.

In orbit I start to unwind after all this excitement. It has been one of the busiest days of my life, but now I am left alone with my thoughts—2 hours per orbit, 48 minutes of which I am behind the moon and unable to talk to anyone. If a census were taken, there'd be 3 billion plus two on one side, and one plus God only knows what on the other. At our preflight press conference I was quizzed extensively about whether I would feel terribly alone or frightened during these intervals, and now I confirm what I suspected before: the answer is absolutely not! I am accustomed to being alone in a flying machine and I like the sensation—the more unusual the surroundings, the better. This is flying solo at its finest. Far from fear, I am enjoying sensations of awareness, satisfaction, confidence, almost exultation.

Columbia is humming along in beautiful shape. I have turned the cockpit lights up bright and removed and stowed the center couch. It is cheery here inside my base camp, waiting for the two explorers down below. I have everything except a fireplace, which wouldn't work too well in weightlessness or 100% oxygen. Columbia's interior is cruciform, like a miniature cathedral. The tunnel through which Neil and Buzz will return is the bell tower, the navigator's station the altar. The main instrument panel spans the two transepts and the nave is where the center couch used to be. It is a familiar, cozy place by now, the fifth day of our expedition, and I fade off into a sleep troubled only by concern for tomorrow's rendezvous. Today's activities have been more successful than I dared hope. We just have to keep the momentum going.

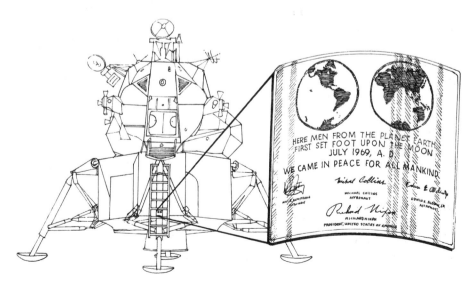

Lunar Module plaque

It is morning again and all goes well down below. As the moment of liftoff nears, I become more and more nervous. Columbia has no landing gear; I cannot help them while they are on the surface. My secret terror for the past 6 months has been to have to leave them there and return to earth alone. If I have to, I will, but . . . it would almost be better not to have that option.

Buzz counts down, and now they are off! For the 7 minutes of their powered ascent, I barely breathe. After much fiddling with my sextant, I finally see Eagle, first as a tiny blinking light in the darkness and then after sunrise as a golden bug gliding through the crater fields below. Reassuringly they grow in my window, steady as a rock as they reduce their closing speed according to a carefully calculated schedule and ease to a stop just 50 feet away. There they are!

I can't see Neil or Buzz, or the 3 billion people on the small blue blob just behind them, but I know they are there—and that's all there are in the entire universe, framed in my window. Now my rendezvous worries can forever be laid to rest. I can throw away the book of emergency rendezvous plans clipped to the front of my pressure suit. God knows we are still a long way from home, but for the first time I feel we are going to carry this whole thing off. I can see it all out my window now, our beautiful home planet and my two compatriots, successfully returned to me. From here on it should be all downhill—for there they are!

My euphoria is briefly interrupted during docking. I have brought us together and made what I thought was soft and graceful contact with Eagle. The two machines are now tenuously hooked by three small capture latches and have to be drawn together to a rigid connection by a retracting mechanism. When I throw the retraction switch, the docile little Lunar Module suddenly veers violently to one side, and I fight with my hand controller to regain a satisfactory alignment between its docking drogue and my probe. Suddenly it veers back and with a loud bang! the rigid connection is made, and all is well again.

As soon as the tunnel between Eagle and Columbia is cleared, Buzz pops through, a big smile on his face. He looks terrific. I grab his head, a hand on each temple. I am going to give him a big smooch on the forehead, as a parent might greet an errant child, but then I get embarrassed and just grab his hand, and then Neil's. We load two precious boxes full of lunar rocks into the zippered fiberglass containers I have prepared for them, and then it is time to dump old Eagle. Personally I am pleased to be rid of it. I don't care care how reliable NASA and Grumman have made it, I have never really trusted that far-fetched contraption. But it has served us well, and I can tell that Neil and Buzz are genuinely touched as we cut it free.

The next link in our daisy chain is another critical one, in NASA jargon called TEI, or trans-earth injection. It should be called ''the get us out of here, we don't want to be a permanent moon satellite'' maneuver. We need to burn our SPS engine for 2½ minutes to break the bond of lunar gravity and put us on a trajectory that should slice through the earth's atmosphere 2½ days from now.

Eagle launching from the Moon

We go through our checklist in meticulous detail, and spend the minutes before ignition worrying about the direction we are pointing. Even though we have checked and double-checked all the numbers we have entered into our computer, still we want to be sure, to see it out the window. Our cue is that the horizon should be nearly on our nose 2 minutes before ignition. "I see a horizon. It looks like we are going forward," I say with a nervous laugh.

"Shades of Gemini," answers Neil, referring to the extreme care we used to take with deorbit burns.

"It is most important that we be going forward," I insist, carried away by nervous laughter now.

Buzz joins in, reciting rocket fundamentals: "Let's see—the motor points this way and the gases escape that way, therefore imparting a thrust that-a-way."

Enough foolishness, let's go home!

Two and a half minutes later, as we swing around the edge of the moon, the earth pops into view, and three very relieved men report the good news to Houston. "Beautiful burn! SPS, I love you, you are a jewel. Whoosh!" We take to the cameras now, tourists getting one last memento of our trip. The moon from this side is full, a golden brown globe glorying in the sunshine. It is an optimistic, cheery view, but all the same, it is wonderful to look out the window and see it shrinking and the tiny earth growing.

The earth as seen from this distance—nearly a quarter of a million miles—is an unforgettable sight.

To begin with, it looks tiny, the size of your thumbnail held at arm's length. It is mostly ocean and clouds, the blue and white dominating the brownish-green of jungles, mountains, and plains. The only land mass that really stands out is the North African desert, especially the oxide-rich, reddish Atlas mountains. Of course Africa may be on the back side of the earth, but wait a few hours and it will swing around into view.

And does the earth glisten in the sunlight! We think a full moon is very bright, but it's a dullard by comparison. In scientific terms, the albedo (reflective power) of the moon is .07; in other words, its surface absorbs 93% of the sunlight and only 7% shines back in our eyes. The albedo of the earth is four times greater, not to mention the fact that the full earth has a reflecting area 13 times that of the full moon. So the earth is a headlight by comparison with the moon. Even the crescent view we are getting now fills our windows with a soft, welcoming light, a beacon beckoning us home.

From a lifetime of prowling its surface, I know that the earth is a huge, solid, rugged place, but from my window now it looks fragile somehow, smooth as a billiard ball, but delicate as a Christmas tree ornament. I wish I had some way of protecting it, of keeping it pristine. It looks so clean and yet it is so dirty, in places at least. The boundary line between a blue and white planet, and one that is gray and tan, *is* fragile. Is the riverbank a delight or an obscenity, a place for diving ducks or greasy truck tires? I cry that the technology that produced this marvelous machine we call Columbia leaves in its wake the detritus of a century of industrial abuse. It need not be that way. We can use technology to cleanse, to repair, to maintain—even as we build, as we spiral out into the universe.

I would like to talk to people about these things, back on the good earth that is now quite close, but it seems more appropriate at this moment to thank those who have made this journey possible. In our last inflight TV broadcast, Neil talks about Jules Verne, and Buzz about ". . . the more symbolic aspects of our mission." I thank "First, the American workmen who put these pieces of machinery together in the factory. Second, the painstaking work done by the various test teams. . . . And finally the people at the Manned Spacecraft Center, both in management, in mission planning, in flight control, and last but not least in crew training. This operation is somewhat like the periscope of a submarine. All you see is the three of us, but beneath the surface are thousands of others and to all those I would like to say thank you very much."

And I will thank them once again when our parachutes open. In the meantime, a storm is brewing in the Pacific and they have moved our landing point 250 miles to the east in search of clear skies and calm seas. I'm not too happy about that, because the new flight plan means that after dipping into the atmosphere initially, we must then extend our trajectory and perform a great soaring arc, quite close to skipping back out into space. But I don't want to land in a thunderstorm either. Houston knows best, I guess.

We each take a motion sickness pill and get ready for entry. We are scheduled to hit our entry corridor at an angle of 6½° below the horizon, at a speed of

36,194 feet per second, nearly 25,000 mph. We are aimed at a spot 80 miles southwest of Hawaii. We jettison our Service Module, our faithful storehouse still half full of fuel and oxygen, and turn around so that our heatshield is leading the way.

Deceleration begins gradually and is heralded by the beginnings of a spectacular light show. We are in the center of a sheath of protoplasm, trailing a comet's tail of ionized particles and heatshield material. The ultimate black of space is gone, replaced by a wispy tunnel of colors: subtle lavenders, light blue-greens, little touches of violet, all surrounding a central core of orange-yellow.

I breathe an extra sigh of relief when we drop below satellite speed, meaning that we don't have enough energy anymore to skip back out of the atmosphere. We are going to land *somewhere* and, according to our computer and our autopilot, in the right place. Over in the left couch, I am watching my instruments like a hawk, ready to take over manually if I see something I don't like. The G forces are squashing us now, like a heavy hand on the chest, but it's only 6½ times earth's gravity, and doesn't last long.

The view out the window is breathtaking. The intensity of illumination has increased dramatically, flooding the cockpit with white light of a startling purity. Our fiery trail has expanded and its edges can no longer be seen. Instead we seem to be in the center of a billion-watt bulb, flooding the predawn Pacific basin with light. As the light show subsides our two drogue chutes deploy. With my eyes fixed in the cockpit I didn't see them release but there they are, flailing back and forth, steadying us enough to allow safe inflation of the three main parachutes. A small jerk, and there they are! Huge, beautiful orange-and-white blossoms of reassurance.

I have bet Neil a beer that when we hit the water we will remain upright and not topple over and float small end down. At the instant of touchdown Buzz must push in a circuit breaker and I must throw a switch to jettison the parachutes, or else the wind will catch in them and drag us upside down.

Splat! We hit like a ton of bricks, and Buzz's hand is jerked away from the circuit breaker panel. By the time he finds the correct breaker again and I throw the switch, it is too late—I can feel us slowly turning over.

Not only have I lost a beer, but we are trapped in here, our escape hatch under water, hanging in our straps for 10 minutes while we pump up small air bags on our sunken nose, changing our center of gravity enough to heave us back upright. While we are waiting to get out we each take another motion sickness pill, not that we feel ill, but at all costs we must not throw up inside the biological isolation garments (BIGs) that the swimmers will throw in to us.

When we open the hatch and receive the BIGs, I go down by the navigation panel to put mine on. It is my first attempt in 8 days to stand upright against gravity. I feel slightly swollen in the feet and lower legs and just a tiny bit light-headed, but even inside this heaving compartment, buffeted by 18-knot winds,

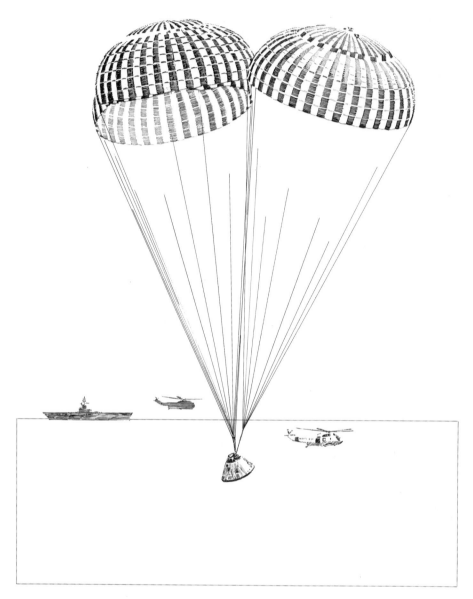

Parachute deployment for Apollo 11 landing on Earth

I feel good—better than I did after Gemini 10. Maybe it's because the last 2 days have been restful or because I have been able to keep my pressure suit off most of the time, or because there is more space in here, or maybe all these factors and others.

Zipped up inside our BIGs we transfer from Columbia into a rubber raft, where the swimmers wash us down with disinfectant to kill any lunar bugs. I don't like to think about what happens to any lunar bugs that drift out of the open hatch. The BIGs have no ventilation, and by the time I have transferred, via a small wire basket, up into a hovering helicopter I am beginning to heat up. By the time the chopper has deposited us on the hanger deck of the USS *Hornet* I am not only burning up but I can't see out of my fogged faceplate. I can hear a brass band, and vaguely perceive a crowd of sailors off to my right. Someone has painted lines on the deck, and I follow them to a low door and pop inside. It's the mobile quarantine facility, a modified house trailer that will be our home for the next 3 days.

After a shave and a shower we three gather at a picture window in one end of the trailer while President Nixon makes a short speech of welcome a few feet outside the glass. "The greatest week in the history of the world since the Creation," he calls it. I'm not sure about that, but certainly the greatest week I have ever experienced, and one I am glad to have behind me.

The USS *Hornet* steams for Pearl Harbor, where we are transferred to a flatbed truck for a jubilant passage along a road thronged with cheering crowds, to Hickam Air Force Base. From there the trailer is loaded onto a C-141 cargo plane for a 7-hour flight to Houston. Another flatbed, another parade—this in the middle of the night—and then we are pulled inside the Lunar Receiving Laboratory (LRL) and the doors lock behind us. Life inside the LRL isn't bad at all. The food is good and we have plenty of room for a change. And plenty of time! For 2 weeks we brief and rebrief NASA management and engineers and future crew members on our flight, putting every tiny detail into our report. Finally, on August 10, 1969, our 3-week quarantine is over, and we are released to the outside world.

As the first men on the moon, Neil, Buzz, and I became a worldwide focus of attention and excitement, but Neil had got it right when he called the lunar landing "One giant leap for mankind." We three were but the apex of a massive effort by thousands and thousands of men and women. Their story, and the spacecraft and space equipment they designed, built, and operated, are the true backdrop to man's exploration of space.

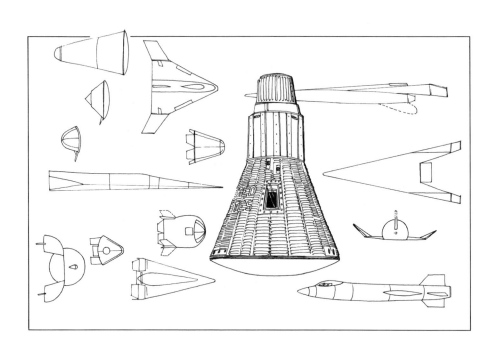

2

An Elegant Solution

On May 2, 1946, a year to the day after Soviet troops captured Berlin, a remarkable document was published in Santa Monica, California. At that time, the Douglas Aircraft Company was the home of Project RAND, the creation of Generals "Hap" Arnold, Curtis LeMay, and other World War II air power advocates. Arnold and his compatriots had wanted to retain the technological momentum of the German V-2 ballistic rocket program, and were intrigued by the future military possibilities of more advanced rocketships. Project RAND (RAND stood simply for Research and Development) later grew into today's prestigious RAND Corporation, but in early 1946 it consisted of four employees in a walled-off section of the second floor of the main Douglas Aircraft plant in Santa Monica.

The 324-page report produced by the prescient combined group of 20 Douglas and RAND engineers is entitled *Preliminary Design of an Experimental World-Circling Spaceship*. Unlike most technical documents, the report makes marvelous reading: in quiet, undertstated terms it outlines a reasoned plan for reaching space. "In this report, we have undertaken a conservative and realistic engineering appraisal of the possibilities of building a spaceship which will circle the earth as a satellite. If a vehicle can be accelerated to a speed of about 17,000 mph and aimed properly, it will revolve on a great circle path above the earth's atmosphere as a new satellite. The centrifugal force will just balance the pull of gravity. Such a vehicle will make a complete circuit of the earth in approximately 1½ hours. . . .

"In making the decision as to whether or not to undertake construction of such a craft now, it is not inappropriate to view our present situation as similar to that in airplanes prior to the flight of the Wright brothers. We can see no more clearly all the utility and implications of spaceships than the Wright brothers could see fleets of B-29s bombing Japan and air transports circling the globe.

"Though the crystal ball is cloudy, two things seem clear:

"1. A satellite vehicle with appropriate instrumentation can be expected to be one of the most potent scientific tools of the twentieth century.

"2. The achievement of a satellite craft by the United States would inflame the imagination of mankind, and would probably produce repercussions in the world comparable to the explosion of the atomic bomb."

The main focus of the RAND report was, naturally enough, on the design of rockets needed to achieve that magical 17,000 mph speed. The RAND analysis compared two different rockets, the first a four-stage design burning alcohol and liquid oxygen, and the second a two-stage machine fueled by liquid hydrogen and liquid oxygen. The alcohol burner built upon the wartime experiences of Wernher von Braun's German Peenemünde team, while liquid hydrogen was an avant-garde effort to squeeze higher theoretical performance from each drop of fuel. The "octane rating" of rocket fuels is called specific impulse, or ISP, in rocket engineering shorthand. ISP is calculated by dividing the thrust of a rocket by its fuel flow. Thrust is measured in pounds, and fuel flow in pounds/second. Therefore, when this division is done, the pounds cancel each other out, leaving the *second* as the unit for measuring rocket prowess. It has always seemed an oddity to me, to be measuring the output of these great roaring machines in terms of a unit of time. At any rate, the ISP of liquid hydrogen is 420 seconds, almost double that of alcohol at 280 seconds. Liquid hydrogen, picked by RAND in 1946, is what launched my compatriots and me to the moon 23 years later, and what fuels the space shuttle today.

Although RAND's four-stage rocket used a more conventional fuel, it was nonetheless designed with some verve and affection. "The vehicle will be divided into four stages, the primary or first stage being nicknamed 'Grandma,' the second stage 'Mother,' the third stage 'Daughter,' and the final satellite vehicle 'Baby.' Baby will carry the payload and intelligence for all stages, in addition to its own fuel, pumps, motor, and guidance, and will comprise between one-fifth and one-fourth of the length of the total vehicle. Daughter and Mother will each carry only fuel, pumps, motors, and controls, being guided by Baby, and will be about the same length as Baby. Grandma will comprise almost half the length of the total vehicle, and will also contain fuel, pumps, motor, and controls, being guided by Baby."

I wonder what vision of the post-1946 future prompted these engineers, these machinery men, to think of their creation in terms of giving birth in mid-flight, each generation passing on its accomplishments to the next, yet guided themselves by an unborn infant?

In addition to achieving orbit, the RAND study also considered other obstacles which Baby had to, and could, overcome along the way: the "method of guiding vehicle on trajectory''; the probability of striking a meteorite; controlling the temperature inside a spacecraft; pitch, yaw, and roll control in orbit; and ''the problem of descent and landing.''

The scope of RAND's report clearly was limited to putting an object, not a man, in orbit, yet—almost apologetically—the authors do bring up the subject. ''Throughout the present design study of a satellite vehicle, it has been assumed that it would be used primarily as an uninhabited scientific laboratory. Later developments could alter its capabilities for use as an instrument of warfare.

''However, it must be confessed that in the back of many of the minds of the men working on this study there lingered the hope that our impartial engineering analysis would bring forth a vehicle not unsuited to human transportation. It was, of course, realized that 500 pounds and 20 cubic feet were insufficient allotment for a man who was to spend many days in the vehicle. However, these values were sufficient to give assurance that livable accommodation could be provided on some future vehicle.''

If this RAND report can be considered a road map to the future of manned spaceflight—and as such, it was an eerily accurate one—then one detour along the way must be attributed to it, and that involves reentry into the atmosphere and the use of a winged vehicle to accomplish what RAND recognized in 1946 would be a very tricky feat.

''An important ultimate goal for any vehicle must be that of carrying human beings with safety. One obstacle which seems to stand in the way in the present case is the great energy stored in the vehicle, a part of which serves to heat the vehicle on descending into the lower atmosphere. The study which follows is an attempt to show the feasibility of lowering the craft without destroying it by fire . . . control of the glide path must be accomplished by aerodynamic means, implying that lifting surfaces must be provided . . . wings of small size will be used for speed control during descent, and for making landings.''

Wings for gliding, wheels for landing. This ''wings and wheels'' assumption plagued designers during the post World War II years. Granted, other organizations besides RAND were pondering the next step in aviation and space, but the technology was coming not from submarines or Buck Rogers, but from the world of aeronautics, where lift was king and drag a dirty word. Some aeronautical engineers spent their entire careers working on airfoil design, making minute adjustments to wing shapes to gain an extra smidgen of lift without increasing drag. One flew to heaven with a high L/D, or lift to drag ratio. The efficient creation of lift, along with more powerful engines, was what aviation had been all about since the Wright brothers, and surely that would continue as we flew higher and faster, even into and out of orbit.

Ballistic missiles, when they came along, had no lift at all (L/D = 0) but then their steep, fiery plunge back into the atmosphere was obviously not fit for humans. It was not until the days of Vostok and Mercury that the notion of using a zero lift spacecraft on a shallow descent trajectory was proved to be the best near-term solution because such a spacecraft is lighter than one with wings and wheels.

The 1946 RAND report estimated that it would require 5 years and $150 million to orbit an "experimental world-circling spaceship." Project Mercury, America's first manned spaceflight program, began in 1958, took 4 years and 8 months, and cost $400 million.

The dozen years between the RAND report and the beginning of Project Mercury represent the golden age of supersonic flight. Edwards Air Force Base, sprawled on a dry lake bed in the Mojave desert 100 miles north of Los Angeles, was the focal point of post World War II flight testing under the direction of NASA's predecessor, the National Advisory Committee for Aeronautics (NACA).

From its creation in 1915, NACA had slowly built up its reputation and its facilities, so that in 1946 it employed nearly 7,000 people at six centers across the country. Its budget ($40 million in 1946) was tiny compared to that of the armed services, but then it did not have to deploy forces worldwide, and spent its money on a smaller scale: wind tunnels and engine test cells. NACA's first published work in 1915 had been a *Report on Behavior of Aeroplanes in Gusts*, and since that time it had climbed the ladder of aerodynamic complexity until, in 1946, it was faced with the airplane's next great technological challenge— the sound barrier.

As an airplane flies faster and faster, the air in front of its nose and wings also moves out of the way with increasing speed. But air cannot move faster than the speed of sound: as the velocity of the plane itself approaches the speed of sound (Mach 1), a wall of compressed air is therefore created in front of it. In 1946 it was not clear whether this wall of air, this speed-of-sound barrier, could be penetrated or not. Certainly it could be by a tiny rifle bullet—but by something large enough to carry a man, with wings to boot?

Theoretical equations indicated that as the Mach number reached 1, drag = infinity. In other words, at precisely the speed of sound, the airplane's drag became infinitely large and it was stopped in its tracks. That was the wall. Scientists joked that beyond the sound barrier there might exist a domain of reverse entropy, in which gas tanks would refill themselves. More disturbing were some cases from World War II of fighter planes diving into the region of compressibility approaching Mach 1 and being unable to recover. Apparently shock waves formed on their relatively thick wings and rendered elevators and ailerons ineffective. Pull as he might on the stick, the pilot continued his terminal dive until the plane disintegrated. Geoffrey de Havilland, son of the renowned British designer, had killed himself in this fashion, practicing for an attempt to reach Mach 1 in one of his father's planes.

In 1946 NACA was eager to investigate these problems using a thin-winged bullet of a rocket plane called the Bell XS-1. Captain Chuck Yeager, at age 24 the Air Force's hottest test pilot, was assigned to Edwards, and on October

14, 1947, in the most historic flight since Orville Wright's, succeeded without difficulty in flying through the sound barrier, to Mach 1.06.

The XS-1's drag had obviously not reached infinity. Clearly supersonic flight (slightly faster than Mach 1) was possible, and perhaps even hypersonic flight (*many* times faster than Mach 1). A whole new golden era for aerodynamics was about to unfold, in which NACA and the military, memories of World War II fresh in their minds, were to create a glistening new stable of supersonic fighters that would ensure a dominant role for U.S. air power. This invincible armada would cause a kind of Pax Aeronautica, not to mention providing fascinating new adventures for those involved in aeronautical research and flight testing.

There was almost a supersonic craze. In a way, Yeager's success in 1947 delayed the United States' entry into space because, like the previous year's RAND report, it caused aerodynamicists to ignore the zero lift space "capsule" and reinforced the notion that winged vehicles could fly faster and faster— supersonic, hypersonic, to orbital velocity if need be. Never mind how much such a machine might weigh.

The Bell XS-1, later called the X-1, was followed by a whole series of more or less successful experimental aircraft. These in turn spawned a new series of Air Force jet fighters that, because of the official numbering system, was called the Century Series. They included the F-100 Super Sabre made by North American; McDonnell's F-101 Voodoo; two delta-winged beauties from Convair, the F-102 and F-106; Lockheed's needle-like F-104 Starfighter; and the sturdy F-105 from Republic; most of these machines could fly as fast as Mach 2 and

Bell X-1

some could reach 90,000 feet in a zoom. When the Soviets put Sputnik I into orbit (October 4, 1957), most of the Century Series fighters were in squadron service and NACA was concentrating on what it hoped would be the *pièce de résistance*, the masterpiece of the X series, the North American X-15.

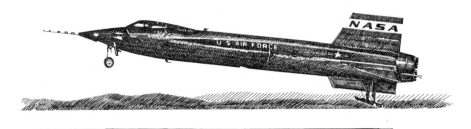

North American X-15

Half airplane, half rocketship, the X-15 was different from its predecessors in that it could fly so high that the air became too thin to use as a control mechanism. In other words, rudders, ailerons, and elevators could be deflected into the slipstream, but the airflow was too weak to cause much response. Full aileron deflection at low altitude would cause a reaction violent enough to bang the pilot's head painfully against the canopy, but at high altitude aerodynamic controls had to be supplemented by a new device, the attitude control thruster. By firing small jets of hot gas from the nose and wing tips of the X-15, the pilot could supplement his aerodynamic control surfaces.

Another difference was the fact that the X-15 flew fast enough to cause frictional heat to be a problem. It was coated black to radiate heat more efficiently, and was constructed of Inconel X, a nickel-chrome alloy capable of sustaining temperatures of 1200°F without losing strength or stiffness. The X-15 was a big, brutal machine, looking as if it had been designed to smash Mach Numbers, not slip by them. It left the sound barrier far behind, the temperature barrier just ahead.

The X-15 began flying at Edwards in 1959, 2 years after Sputnik I. Yet there was no appearance of a contest between the two. Sputnik I was a 184-pound cannonball, the X-15 a 30,000-pound flying machine. By the time I reached Edwards in 1960, the X-15 flight test program was going full blast, and by 1962, it had reached Mach 6 and a 60-mile altitude, its design limits. 1962 was also the year the 3000-pound Mercury capsule reached orbit.

Five years earlier, on October 15, 1957, at NACA's Ames Aeronautical Laboratory in Northern California, some of the country's leading aerodynamicists gathered to consider ''Round Three.'' Round One had been the historic X-1; Round Two would be the Mach 6 X-15; but beyond that—what shape must a vehicle have, to fly 12 or even 18 times the speed of sound? How could the tremendous aerodynamic heating be handled?

Maxime A. Faget, a NACA aerodynamicist stationed at Langley, Virginia, was one of the engineers called to Ames. He was making the standard NACA coast-to-coast trip in a World War II surplus transport, the DC-3. It was a 2-day trip, with a stopover at El Paso, Texas. Poker was the most popular way of making the hours pass a little faster. The plane was unpressurized but had to climb to 15,000 feet or so to clear the Rocky Mountains. At this altitude the oxygen content of the passengers' blood was reduced, and they became mildly hypoxic. Max could tell when this was happening by watching the fingernails of his fellow poker players. When they turned blue, Max would make his move: he began to bluff outrageously and usually successfully.

Max Faget was also a canny engineer. Small, wiry, with a Louisiana Cajun drawl, Max had served as a submarine officer following graduation from Louisiana State University, and then joined Robert R. Gilruth's newly formed Pilotless Aircraft Research Division at Langley, Virginia. Gilruth was pioneering new ways of measuring a test model's aerodynamic characteristics at high speed. Dissatisfied with the limitations imposed by wind tunnels, Gilruth and Faget dropped lead projectiles from aircraft at high altitudes, and put winged shapes on the front end of small rockets—anything that might accelerate an airfoil to higher and higher speeds—and then measured, via radio signals sent to a ground station, its lift and drag. These simple radio signals later grew into what was called telemetry, wherein thousands of bits of information were relayed from spacecraft to ground stations.

Max was accustomed to the tedious trip to California, but this time it was different. Just the week before, the Russians had launched Sputnik I, and Max was not sure how that might affect Round Three. The conference would be working toward a big jump in performance beyond the X-15—but all the way to orbit? As the DC-3 droned along, Max interrupted his poker game to wonder about some of the options he had been studying. . . . Perhaps something that could skip in and out of the atmosphere? Perhaps something shaped like a half cone on the top and flat on the bottom?

Sputnik I's influence on Round Three was mixed. Some participants pretended Sputnik I just hadn't happened. With this business-as-usual contingent, the conversation stuck to lift-to-drag ratios, good old L/D. "That's a nice high lift, low drag shape, Harold, but watch out if the shock wave reaches the boundary layer. It'll heat up like an iron skillet." Others thought that L/D had been made academic by Sputnik I. It might be a zero lift cannonball, but there it was, orbiting the earth every 90 minutes. You couldn't say at what Mach number it was traveling, because Mach was meaningless above the atmosphere, but there it was, Harold, going around once every hour and a half.

Two of Ames's heavyweights tried out conflicting approaches. Alfred J. Eggers, Jr. called his hypersonic glider the M-1 (although some others, like Faget, remember it as "the half baked potato"). Years later, such rounded half cone designs would be called lifting bodies. The idea behind the Potato was that although it was not a high L/D vehicle, it could generate enough lift to allow the pilot to pick out a landing spot and glide to it, and—more important—the lift could be used to reduce the deceleration caused by entering the atmosphere from very high speed and altitude—or from orbit. Al Eggers was worried about how many Gs a pilot could safely endure.

Harry Julian Allen, known as Harvey, had a different way of looking at the situation. He reasoned that the heat generated at hypersonic speed was so great that conventional aerodynamic thinking had to be abandoned. Specifically, he abandoned streamlining and offered a shape quite offensive to those who had spent their careers refining the thin wing and the needle nose. Harvey's shape was blunt to an extreme, looking like a World War I doughboy's helmet flung into the wind, not sideways like a frisbee, but head on. The key to Harvey's design was that it produced high drag.

To aerodynamicists, weight and drag had always been considered enemies, the foils of lift and thrust. Harvey made drag an ally, the trick being that if such a blunt shape traveled at high enough speed, a strong bow shock wave would be formed in the air ahead of the vehicle. Air temperature in the vicinity of the shock wave might reach several thousand degrees, but most of the heat would remain in the air and not touch the vehicle itself. This "Man in a can" was Harvey Allen's solution to RAND's lament about "the feasibility of lowering the craft without destroying it by fire."

Faget believed Allen 100%. On the DC-3 back to Langley, he became really excited. "By God," he thought, "things are going to be different now. Here's a chance to think things no one has thought about before, to put man in orbit. To hell with all this high class aerodynamic theory." Within a couple of days back at Langley he had six people working on various shapes. Very shortly afterward he came up with the basic shape of the Mercury capsule. His design considered both the total heat load and the peak heating rate of the capsule during atmosphere entry, and sought to minimize them. "I knew they couldn't beat me if weight turned out to be the issue," Max told me. And weight was the issue. With Al Eggers' wings had to come wheels, and all that added a great deal of weight.

"Capsule," by the way, seems to have slipped unnoticed into the engineering vocabulary at this time, just as aeronautics turned into astronautics. Certainly the new Mercury shape looked more like a thimble or a trash basket than an airplane, so perhaps capsule was an apt compromise. But when I entered the space program a few years later, we astronauts detested the word and always used "spacecraft" instead. A capsule was something you swallowed.

While Faget was working on the shape of Mercury, Bob Gilruth was wrestling with a different set of problems. By that time Gilruth was already an old NACA hand. He had been assigned to Langley, in tidewater Virginia, in 1936, and in 1941 he had prepared a seminal report *Requirements for Satisfactory Flying Qualities of Airplanes*. His specialty was pleasing pilots by explaining in engineering parlance what made an airplane easy and pleasant to fly.

During World War II the British sent a delegation to Langley to discuss his work, and "I got famous at that," he told me. Bob also did a great deal of airfoil testing, and was influential in making sure that the wings of Chuck Yeager's XS-1 were thin enough to reach supersonic speed. Some of the World War II fighters had such thick wings that at high speeds, as shock waves formed on them, the pilot found increasingly heavier stick forces. The P-47 was so bad in this respect that Gilruth once hired a pilot who was also a weight lifter to do some dive tests.

Robert Gilruth

Despite his close working relationship with pilots, Gilruth became head of a group called the Pilotless Aircraft Research Division after World War II. Using facilities at Langley and nearby Wallops Island, Gilruth pioneered new techniques for testing airfoils at high speeds, and it was to help in this work that he had hired Max Faget.

Max was an ex-submarine officer fresh from the war, and he and a college buddy, Guy Thibodeaux, traveling across the country, were job hunting together. By the time they got to Langley they were running out of clothes. Both the same size, they traded with each other, and ended up at their Gilruth interviews looking like twins with Hawaiian sport shirts, khaki pants, and sandals without socks. "Gilruth must have been hard up because we sure didn't look impressive," Faget recalls. But Gilruth's judgment was impressive, for Thibodeaux turned out to be Langley's rocket engine expert, and Faget the premier spacecraft designer of his time.

By 1958, Gilruth had been promoted from the Pilotless Aircraft Research Division to Assistant Director of Langley, and was about to change jobs once again. In the aftermath of Sputnik I, the military, scientific, and technical communities of

Max Faget

the United States were in a state of turmoil, and the Eisenhower administration was groping with the problem of creating a unified organization to duplicate, and then to surpass, what the Soviets had done.

The Army offered a suborbital shot, "Project Adam," which NACA Director

Hugh Dryden described as having "about the same technical value as the circus stunt of shooting a young lady from a cannon." The Navy weighed in with a complicated inflatable glider called MER I. And the Air Force, miffed at the intrusion by its sister services into what it considered its domain, proposed MISS—"Man in Space Soonest," a ballistic capsule that would first carry instruments and then a man.

NACA cooperated with the Air Force on the design of such a vehicle, primarily using Faget's ideas. At the same time, aerospace contractors such as the McDonnell Aircraft Corporation of St. Louis were keeping their ears to the ground and their designers busy with their scratch pads.

In January 1958, the Air Force pulled together an interesting group of NACA and industry people at Dayton, Ohio, to consider the various possibilities for getting a man into space quickly. The NACA attendees discussed two designs: a zero lift capsule in the form of a curved heatshield with a tapered afterbody; and a winged lifting body. The industry proposals generally acknowledged the advantages, from the viewpoint of weight and simplicity, of a zero lift sphere or cone. But more than half a century of airplane heritage could not be erased, and several ideas were of the super-airplane type.

North American, for example, proposed a modified X-15 rocket research aircraft that would enter the atmosphere off the west coast of Florida, whereupon the pilot would eject and descend by parachute. Republic's glider also was designed to crash while the pilot parachuted to safety. Northrop liked a recoverable winged vehicle called the Dyna-Soar. The Convair division of General Dynamics wanted to hold out for a manned space station. Avco's design was most imaginative: a sphere to which was attached a stainless steel parachute shaped like a badminton shuttlecock, the diameter of which could be varied by a bellows operating on compressed air. Bell, Ford, Goodyear, Lockheed, and Martin had zero lift cones or spheres. McDonnell noted that a minimum zero lift vehicle would weigh around 2,400 pounds and experience a deceleration of from 9 to 14 Gs. A winged vehicle would undergo less than 2 Gs during entry but would weigh about 6,000 pounds.

McDonnell's minimum vehicle looked a lot like Faget's design, and for good reason. Lawrence Michael Weeks, head of McDonnell's advanced design group, had been keenly interested in the Air Force's "Man in Space Soonest" from its inception. For technical help he had turned to NACA and its Langley wind tunnels and—who else?—Max Faget. Mike Weeks began with a sphere, but then decided it would be too heavy for the Atlas booster rocket, so he switched over to Faget's notion of a protruding heatshield with a conical afterbody just large enough to house a man. (The Soviets, who did not have such tight weight limits, opted for the entire sphere.) Mike added to his design team until it consisted of "50 of the best engineers we had." They were all airplane people, none of whom had ever worked on anything faster than the F-4 fighter ("We got it going Mach 2.7 and almost melted the canopy off").

The founder of the company, James S. McDonnell, was not convinced Mike was on the right track. Every so often "Mr. Mac," as he was known around St. Louis, would bellow at his Engineering Vice President: "Tell Mike to cease and desist," but somehow he was always talked out of it. Mike persisted and, as he did, the bill for the design of his embryonic spacecraft soared. Mr. Mac (who had been known to pinch a penny) got into the habit of calling Mike each Friday afternoon for a status report, and when his investment reached $353,000 he blew his stack. "Mike," he stormed, "prepare a memo for Monday's Board meeting explaining why we have wasted $353,000 on basketballs in space," using one of President Eisenhower's favorite analogies.

Mr. Mac was also incredulous that anyone in his right mind would put a man on top of something as fragile and unreliable as an Atlas ICBM (intercontinental ballistic missile). Not sure that he was getting the straight story from his people, he called Convair in San Diego to hear at firsthand that the Atlas would not collapse under the weight of Mike's capsule.

While the technical wheels were churning throughout the industry, the political process was also grinding away. Each of the military services was lobbying hard for a preeminent role in space, and many key members of Congress supported the military because of their own nervousness about new weapons that could skip over the nation's traditional barriers, the Atlantic and Pacific Oceans.

In the wake of Sputnik I, President Eisenhower appointed Massachusetts Institute of Technology President James R. Killian, Jr. a Special Assistant to the President for Science and Technology. Eisenhower gave Killian and his Science Advisory Committee the job of organizing the country's space effort. Not convinced that basketball games in space should be militarized, Eisenhower also made sure that Killian's group looked at other government agencies that might provide space leadership.

The Atomic Energy Commission was considered, because of its experience in managing large, highly technical projects. And, of course, there was the National Advisory Committee for Aeronautics, which had been responsible for the government's aeronautical research since 1915. Two words are especially important here: "research" and "committee." Most senior NACA people were accustomed to doing research only—that is to say, they turned their promising ideas over to others for development and production. They were uncomfortable with the idea of running factories, riding herd on costs and schedules, or solving complex production problems. Some Young Turks within NACA were salivating at the chance to take on the whole space job, but generally NACA was regarded as a "research only" organization.

NACA was also a committee and worked by consensus. Although this system seemed to have served the United States well over the years, it was not appealing to bureaucrats, especially those in the Budget Bureau. What they wanted was an agency headed by one person, with one firm hand at the helm, one member of the President's team who would be immediately responsive to the administration, without having to check with something as amorphous as a committee. Yet everyone had to admit NACA was good; NACA was where the engineering brains were.

These shifting currents, acknowledging Eisenhower's preference for civilian control while rejecting the committee structure, resulted in the creation of NASA, the National Aeronautics and Space Administration, or Space Agency as it is informally called. After a great deal of backing and forthing within the Congress, especially within Lyndon B. Johnson's Senate Special Committee on Space and Astronautics, the Space Act of 1958 was passed, and signed by the President on July 29. The Act did not delineate a division of responsibilities between NASA and the military, but it was clear that Eisenhower wanted NASA to manage the manned space program, and the first step was to be what came to be called Project Mercury.

Gilruth and his people at Langley had not waited for the creation of NASA. Early in 1958, Gilruth had put Faget and Caldwell C. Johnson to work on refining Faget's zero lift design. What emerged was the Mercury capsule: a gently rounded, blunt face containing the heatshield, and behind it a truncated cone with a small cylindrical aft section. Much work went into optimizing the radius of curvature of the heatshield, the final design having a diameter of 80 inches and a radius of curvature of 120 inches, the 1:1.5 ratio chosen because of its superior heat rejection capabilities. During the spring and summer of 1958, Caldwell, who started his engineering career as a draftsman, drew and redrew the capsule, while the politicians in Washington drafted and redrafted the Space Act.

NASA's first administrator was T. Keith Glennan, president of the Case Institute of Technology, a former member of the Atomic Energy Commission, and a loyal Eisenhower Republican. Presented with the results of Faget's work, 1 year and 3 days after Sputnik I, he responded enthusiastically, "All right. Let's get on with it." To get on with it, NASA created the Space Task Group (STG) at Langley, and put Bob Gilruth in charge, with a cadre of 35 people. On December 17, 1958 (55 years to the day after Orville Wright's first flight), Keith Glennan publicly announced that the new venture, to put a man in space as soon as possible, would be called Project Mercury.

Thus began one of the greatest and most successful engineering adventures of all time, one that saw the basic Mercury capsule evolve into the Gemini spacecraft and the Apollo Command Module that made nine lunar round-trips.

With Project Mercury every airplane company worth its salt sensed the beginning of a whole new ball game, with very different rules. No longer would air, or the lack of it, determine where a machine might operate. High above the atmosphere, a vehicle moving fast enough could produce enough centrifugal force to balance the earth's gravitational pull, and a stable orbit would ensue, as RAND had forecast and Sputnik I had made so painfully clear. In this strange

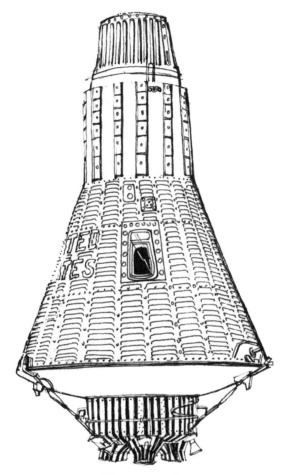

Mercury capsule

new region called space, or the condition called weightlessness, man would soon test his own ability to function. There were questions of radiation, meteoroids, noise, vibration, acceleration, and the vacuum of space. Some data already existed, but as is usual with scientific research, each new answer spawned its own series of questions.

First, and probably most controversial, was the effect of weightlessness on the human body. High performance jet fighters flying a parabolic arc could duplicate weightlessness for half a minute or so, but not too much could be learned about physiological abnormalities in such a short interval. Yet experimenters made the best of it, eating, drinking, and even checking bladder function. One Air Force flight surgeon flew more that 4,000 of these "Keplerian" trajectories without ill effects.

Nevertheless, many experts feared that some body organs required sustained gravity, and that the consequences could be fatal if they were deprived of it for very long. The digestive function, for instance, might be interrupted if nutrients simply floated through the stomach and bowel. More seriously, the heart muscle—like a cavitating pump—might churn away without producing significant blood flow. Neurological side effects were also suspected, because in earth's gravity the brain receives reassuring signals from pressure points in muscle, nerve, and organ tissue, and from the labyrinth of the inner ear. In weightlessness these signals could disappear or contradict each other, so that the brain—unable to decipher the messages—might simply shut down and cease functioning.

NACA, devoting itself exclusively to flight technology, had left biomedical research primarily to the military, and so was not much help in answering these questions. Starting with V-2 firings from White Sands after World War II, the Air Force had used animals (primarily mice and monkeys) as sounding rocket passengers. Again, there appeared to be no ill effects, but no military flight surgeon in his right mind was going to stand up and swear that sustained weightlessness was safe and that Project Mercury should proceed posthaste. Far from it. Before committing a man to Mercury, the Air Force medics wanted 15 or more launches of primates and small biological payloads.

If not enough gravity was a problem, so was too much. Orbital flight was one long freefall, but it was preceded by a jolting rocket ride and, even worse, ended with a flesh-flattening plunge back into the atmosphere and a final splat of a landing. Acceleration going up, deceleration coming down. Aboard Mercury a body had no way of telling the difference because the capsule ascended small end first and descended with the blunt heatshield forward. In technical terms, the acceleration was transverse to the body, or, as pilots put it, "eyeballs in." In airplanes, tight turns cause blood to rush toward the pilot's feet, so this G direction was called "eyeballs down." The Air Force knew from high performance fighters that sustained eyeballs down G levels of 6 or 7 were barely tolerable, but that the body could easily handle twice that eyeballs in. Unfortunately, some of the trajectories onto which the zero lift Mercury might stray ended with 20 Gs, which might cause those eyeballs to stay "in" permanently. That was the price Max Faget had to pay for designing a weight-saving, zero lift capsule, and that worried him plenty.

Fortunately, the acceleration of a spaceflight is a lot easier to duplicate on the ground than weightlessness. All it takes is a centrifuge, with a motor powerful enough to swing a large gondola, mounted on a long arm, around in circles at high speeds.

The Germans experimented with centrifuges during World War II, and the U.S. Navy operates one with a 50-foot arm at Johnsville, Pennsylvania. When I was a test pilot at Edwards we used to be called on occasionally to ride this centrifuge, or "wheel," as we called it. Nothing I can think of in a pilot's repertoire is more miserable than the wheel. As the arm picks up speed, you are forced back in the seat or couch. Pinned there, unable to move arms or legs, you find all the breath forced out of your lungs and your vision begins to narrow and darken. By the time you reach 15 Gs, there is a sharp pain beneath the breastbone. You can't inhale, and you can barely make out the instrument panel. As John Glenn wrote, "at 16 Gs it took just about every bit of strength and technique you could muster to retain consciousness."

When you emerge from the gondola you experience extreme dizziness if you turn your head sideways, and for a day afterward you have a world-class hangover. July 29, 1958, was therefore a banner day for the Mercury program. Navy Lieutenant Carter Collins sustained 20 Gs without injury, riding in a prototype Mercury couch on the Johnsville wheel. At Langley, Max Faget breathed a sigh of relief.

Just 6 months before this, the first U.S. satellite, Explorer 1, had radioed back strange news: its Geiger counters reported unexpectedly high radiation, coming from what became known as the Van Allen belt. Consisting of charged particles trapped by the earth's magnetic field, there are actually two Van Allen belts. The inner belt, which consists of protons, is the more dangerous of the two. There are also very high energy cosmic rays coming from outside the solar system, and our own sun erupts periodically, sending deadly radiation streaming toward earth. On the ground, we are shielded from all this radiation by the atmosphere, but above it a human is protected only by the spacecraft's structure, which tends to be very thin because of weight constraints.

Then there are meteoroids. We see these projectiles as meteors, burning up as they penetrate the atmosphere, or find them on the ground, where they are called meteorites. But before they become meteors or meteorites, these meteoroids zing through space at very high speeds (up to thousands of miles per hour), and if they are big enough they can puncture a spacecraft's hull.

The vacuum of space works both for and against the designer. It's nice to be able to extend an antenna or an optical device without any slipstream to rip it off. On the other hand, the pressure shell of the spacecraft has to be more reliable than the cabin of an airplane because a leak into space could mean sudden death unless the occupant is wearing a pressure suit. Human blood boils above 63,000 feet pressure altitude.

Noise and vibration were also problems. Communications were considered vital, especially during the noisy first seconds of flight when the Atlas' huge rocket motors were blasting away, their exhaust gases bouncing back off the concrete launch pad. Vibration could not only interfere with vision, but at certain frequencies a human's abdominal organs could resonate, amplifying the vibrations to the point of ripping themselves loose.

As if the hazards of weightlessness, acceleration, radiation, meteoroids, vacuum, noise, and vibration were not enough, the medics raised one final issue. How would spaceflight affect the psyche? Sealing people in isolation chambers on the ground produced mixed results. Some seemed unchanged, while others exhibited surprisingly strong symptoms of hostility, or became unable to sustain their proficiency at even simple tasks. Major Charles Berry, an Air Force flight surgeon assigned to Project Mercury, summed up one school of thought: "The psychological problems presented by the exposure of man to an isolated, uncomfortable void seem to be more formidable than the physiological . . . ."

With due consideration for all these sources of potential catastrophe, Gilruth and his Space Task Group proceeded swiftly. In the back of their minds they knew the Soviets would soon be following up Sputnik with a manned flight. Little by little, Faget's wingless design had gained favor within NACA because it was light and simple, and by the time the Space Act of 1958 was passed, Al Eggers' half baked potato, its closest competitor, was really out of the running. Competition was still coming from the Army, Navy, and Air Force, with their Adam, MER I, and MISS, but when Eisenhower gave the manned program to NASA instead of the Defense Department, Faget's design became the only viable candidate. Max had won the design competition almost by default, or by "silent consensus," as he described the process to me.

Max Faget and Chuck Mathews, a Langley aeronautical engineer, who was later to head the Gemini program, had by now fleshed out the design and written a specification document upon which a contractor might bid. Thus Harvey Allen's concept had evolved into Max Faget's configuration, a machine now awaiting its transformation from paper into titanium, aluminum, and steel. When NASA Administrator Keith Glennan said "Let's get on with it," Bob Gilruth's Space Task Group was ready.

On November 7, 1958, a briefing was held at Langley for 40 industrial firms, the list comprising virtually all the U.S. corporations that built airplanes or their principal components. A week later, Mathews' specification was mailed to 20 of these companies that had indicated a desire to be considered as bidders. On December 11, twelve of the solicited firms and one unsolicited firm, Winzen, submitted bids. A NASA Source Selection Board worked hard over the Christmas holidays and completed its evaluation on January 9, 1959. It selected the McDonnell Aircraft Corporation of St. Louis. The losers were Avco, Chance-Vought, Convair, Douglas, Grumman, Lockheed, Martin, North American, Northrop, Republic, and Winzen. They had bid on something called "the manned satellite project," but by the time McDonnell was picked, the Mercury had been officially announced.

The evaluation had boiled down to a two-way technical competition between Grumman and McDonnell, but Grumman was thought to be already overloaded with Navy business, while McDonnell had more manpower to put to work immediately. In fact, Mike Weeks and his team of 50 had by then been working on Faget's basic shape for a year. When McDonnell won the award, Mike stayed in his advanced design job, and the task of building the Mercury fell to John F. Yardley.

The first problem confronting Yardley and his NASA customer was the Atlas booster. The Atlas had had a checkered history going back to 1946. Canceled once, underfunded badly until the mid-1950s, the Atlas had been designed as an intercontinental ballistic missile capable of carrying a thermonuclear warhead. The possibility of its carrying a man had not been considered in its original design, and the extra margin of safety built into manned equipment was obviously lacking.

Fighting for every ounce of weight, its designers had used what was called a "gasbag" structural scheme in which paper-thin stainless steel sections were made rigid by inflating the rocket with pressurized helium. Without the pressure, the rocket sagged. Mr. Mac's phone call to Convair inquiring about the Atlas' ability to handle Mercury's weight without collapsing had not been facetious.

An Atlas without a payload consisted 98% of fuel, by weight. How heavy a payload could it safely deliver to orbit, and how much would the Mercury capsule weigh? These two key questions had to be answered with some precision and quickly, unless one wanted to wait for a second generation ICBM called the Titan, or a souped-up version of the intermediate range Thor. To have any chance of beating the Soviets, the Atlas, for better or worse, was it.

When McDonnell had been working on Man in Space Soonest, for the Air Force, Mike Weeks talked in terms of a "1-ton" spacecraft. When Chuck Mathews wrote the Mercury specification, the target weight was 2,700 pounds. Preliminary McDonnell calculations put the capsule's weight at 2,400 pounds, plus or minus 25%. Now minus 25%, or 1,800 pounds, was fine for the Atlas, which in late 1958 could guarantee only 2,000 pounds to orbit. But plus 25%, 3,000 pounds, would be too much, and *plus* 25% was a lot more likely than *minus* because early designs almost invariably contain errors of omission. "People like Max and myself never seem to understand that extra things get imposed on the system," Caldwell Johnson told me. Yardley and company could not afford any extras.

While NASA and McDonnell were struggling with the capsule's weight, the Air Force and Convair were trying equally hard to increase the Atlas's performance. In a prototype version, each of its two Rocketdyne booster engines produced 135,000 pounds of thrust, and gradually this number crept up to 150,000 pounds and slightly beyond as the propulsion engineers made a series of esoteric plumbing changes. They also applied tricks such as supercooling the already chilly (minus 297°F) liquid oxygen, thereby increasing the weight of oxidizer that the tank could hold by 1,000 pounds. And they began an extensive weight reduction program, a reexamination of every component, to shave off a pound here or even an ounce there.

Calculations were made over and over again on the effects of wind, temperature, guidance accuracy, and anything else that might add to or detract from performance. The perigee, the point nearest to the earth, of the planned Mercury orbit was reduced from 107 to 103 miles, which made the Atlas' job a bit easier. One hundred three miles was as low as the designers dared to go, because below that the atmosphere density was suspected to be sufficient to drag the capsule back to earth prematurely.

Reducing weight and increasing performance, although especially critical in the case of the Atlas, were familiar problems to the Space Task Group engineers. Most new aircraft designs suffered through a similar process. The thing about the Atlas that was different, that was beyond their ken, that was absolutely hair-raising, was its reliability record.

It was horrible. A few months before the formation of the Space Task Group, Bob Gilruth had testified before a committee of the Congress that "The Atlas . . . has enough performance . . . and the guidance system is accurate enough, but there is the matter of reliability. You don't want to put a man in a device unless it has every good chance of working every time." Gilruth felt that these problems could be worked out with more test firing. "Reliability is something that comes with practice."

But as 1959 began, this reliability did not seem to be coming. Seven out of eight sequential launches were failures. Sometimes the Atlas staggered off course, sometimes it blew sky high, but in any event it was not like any other machine a man had ridden so far. A year earlier Max Faget had calculated that the Atlas' record for reaching orbit safely was 40%; now it seemed even worse than that, yet at least 80% reliability was required before risking a man atop this fiery, fragile gasbag.

Unlike an airplane test program, there was no way to ease into the Atlas. With an airplane, you first did some taxi tests, then flew briefly with the gear and flaps down, then increased speed and altitude step by small step in a methodical progression. With a big rocket it was all or nothing. Strap that baby on and—zoom—you were off, for better or worse, to maximum speed and altitude, to the four corners of the flight performance envelope, as the engineers called it. It was quite a cultural shock for the staid aeronautical engineers of the Space Task Group to meet the Atlas.

During 1959 and 1960, as the Mercury capsule neared production, the Atlas underwent several changes as part of an upgrading program to make it reliable enough to carry man. A fiberglass shield was fitted over the dome of the liquid oxygen tank, to prevent the Mercury capsule's own rocket motors from burning through the tank when the Mercury separated from the Atlas. An older, more reliable type of engine valve was substituted, special lightweight telemetry was used, and the engine start procedure was modified to reduce structural stress.

The major change, however, was the addition of an abort sensing system. Not happy about calling it ASS, the Air Force officially named it ASIS, or Abort Sensing and Implementation System. NASA people suspected this was to be read "as is"; in other words, the Air Force was making the point that NASA was buying the big bird as is, basically a bare Atlas with a few embellishments to make it "man-rated." Gilruth was counting on experience to be the great teacher, to raise launch reliability to an acceptable number.

Aside from its reliability, there was another drawback to Atlas in the eyes of the Mercury managers: at over $3 million apiece, it was too damned expensive.

(How times have changed!) Also, in an attempt to sneak up on orbital flight, NASA was planning, airplane-like, some preliminary tests of Mercury and its components at less than orbital speeds and altitudes. Some of these would use empty capsules, while others would carry chimpanzees. The Redstone rocket, a product of Wernher von Braun's Army team in Huntsville, Alabama, had been in test since 1953, and by 1958 had compiled a safety record far better than the Atlas'. It could lift the Mercury capsule, if it didn't get any heavier, to an altitude of 100 miles, but it didn't have the extra power to propel it to

Redstone rocket

17,500 mph, the orbital velocity at that altitude. For that the Atlas was required. However, the Space Task Group engineers suggested half a dozen preliminary Redstone flights as a less expensive alternative to the Atlas.

In a further effort to reduce costs, Faget created Little Joe, a booster that consisted of four solid rockets clustered together in an awkward-looking package. The rockets, modified Sergeants, had been in use at Wallops Island for years, so the basic elements of a launch system already existed. Like the Redstone, Little Joe was unable to propel the Mercury to orbit, but it could be useful in testing components such as the capsule escape system and the parachutes. At one time, Faget thought that he could make Little Joe reliable enough to carry a man, but that plan was abandoned and that job was left to Redstone and Atlas.

By early 1959, Bob Gilruth's Space Task Group at Langley had become swamped with work, and the organization was expanded rapidly. The original contingent of 25 engineers had been hand-picked by Gilruth. (''I knew who the good guys were.'') But Floyd Thompson, Langley's Director, balked at continuing this system. From now on, he decreed, he would choose one member of the team for every one Gilruth picked. ''We never knew for sure what list we were on,'' quipped Caldwell Johnson.

In addition to robbing Langley of talented engineers, Gilruth added outsiders to his team. By a lucky coincidence, the Canadian government picked this time to cancel the Avro CF-105 Arrow, a supersonic interceptor, and Gilruth acquired 25 experienced Canadian engineers. Personnel were also transferred from other NASA centers, most visible among them Walter C. Williams, the chief of the High Speed Flight Station at Edwards, California. These additions made Gilruth's organization not only larger but better balanced. Langley's strength had always been research, but the production and test phases had traditionally been left to others. Now the Space Task Group, with the responsibility for all aspects of Project Mercury, was delighted to welcome engineers who were experts in developing, manufacturing, and operating high performance flying machines.

And the Mercury would certainly be a high performance vehicle by anybody's standard. Although accelerated into orbit by the Atlas, from then on it would be on its own, and—with a total weight about equal to that of a compact car—it would have to include myriads of features to sustain its fragile human occupant in an environment far harsher than that of an automobile or even an aircraft.

Automobiles and airplanes traditionally rely on steel and aluminum, respectively, for their basic structure, but steel is too heavy and aluminum too soft at high temperatures for most space applications. For its basic structural shell the Mercury designers turned to titanium, a relatively brittle metal with a high strength to weight ratio. Titanium had been used before in aircraft, but on Mercury it had to be much thinner, which made welding it a critical operation. McDonnell learned to weld titanium only .01 inches thick in an inert argon atmosphere. The titanium hull was blanketed by fiberglass insulation, which in turn was covered with blackened heat radiating shingles. The shingles on the conical side walls of the capsule were made of Rene 41, a nickel steel alloy, while those on the higher temperature cylindrical afterbody were of beryllium.

The material selected for the heatshield was a subject of extended debate. Indeed it was not so much the material as the method of heat rejection that was

Little Joe rocket

Heatshield ablation

the fundamental question. There were two ways of handling the frictional heat generated by the descent into the atmosphere: it could be sopped up (the ''heat sink'' principle), or it could be ablated. Ablation, a word used in the study of meteorites, is a process which intentionally causes part of the spacecraft to be consumed by fire. As the outer surface of an ablative heatshield vaporizes, heat is carried away with the lost material. The ablating surface does not simply disappear, but chars, creating a protective carbon coating that assists by radiating additional heat. Ablation showed promise as a lighter solution, but to some engineers it seemed an untidy process, and it was a risky business to calculate the shield's thickness precisely.

Mercury was expected to encounter temperatures as high as 3500°F during entry, and fragile fiberglass and phenolic resin, the materials of ablation, seemed much less predictable in that environment than a good solid copper or beryllium heat sink. Hugh Dryden, who had become Glennan's deputy, favored a beryllium sink, and Mercury started off in that direction, but Gilruth and Faget were able to gather enough data to convince him that an ablative shield was superior, especially in ease of manufacture, and the decision was reversed in time to provide ablative heatshields for the orbital flights.

The shape of the Mercury was the logical outcome of Faget's training in aerodynamics, tempered by the weight-carrying capacity of the Atlas. Given shape and weight, the capsule's dimensions followed easily, the two most important determinants being the 6-foot diameter of the Atlas and the length and breadth of the human body. Looking back on it nearly 30 years later, Mike Weeks finds it "amazing that we would put a guy in that confined a space."

Federal prison specifications require a minimum cell space of 475 cubic feet per inmate; the Mercury has a habitable volume of 40 cubic feet, about the same size as a coffin. Remembering the 1-ton early payload days, Gilruth recalls, "I got serious suggestions from industry that we should get a pilot without legs." A space age Douglas Bader was not to be, however, as the Mercury astronaut specification allowed a height of 5 foot 11 inches. The overall length of the Mercury capsule was 9½ feet, and it would have fitted cozily into the engine bell of one of the shuttle's three main engines.

Of all the equipment crammed into the Mercury capsule, none was more vital than the environmental control system (ECS). The ECS controlled the atmosphere inside the capsule. In selecting this atmosphere, designers of the ECS had to think about hypoxia, the bends, toxicity, equipment weight, and fire. Their choice, pure oxygen, at a pressure of 5 pounds per square inch (psi), represents a typical design compromise.

On earth we breathe air at a pressure of 15 psi, so the first question was whether pressure in the capsule could be safely reduced to one-third of surface terrestrial atmosphere. Air is composed of 80% nitrogen and 20% oxygen, so that on earth the partial pressure of oxygen is only one-fifth of the total, or 3 psi. Therefore, there are more oxygen molecules in the lungs at 5 psi, 100% oxygen, than there are at 15 psi, 20% oxygen, and hypoxia is no problem. In addition to oxygenating the blood, the ECS must exert sufficient pressure on the body to prevent the bends. The bends occur when nitrogen, under reduced pressure, comes out of solution in body tissues and forms gas bubbles. These bubbles can cause pain, especially in the joints, and can even affect the central nervous system with fatal results. So pressure inside a spacecraft cabin must be sufficient to prevent nitrogen from gasifying, and, aside from the nitrogen problem, to prevent body fluids from vaporizing, which occurs at cabin altitudes greater than 63,000 feet, where the pressure drops below 1 psi.

Breathing easily, with no gas in his blood, the astronaut next complicates the designer's life by exhaling carbon dioxide. Somehow it must be removed or it can build up in the cabin, causing hyperventilation and eventual loss of consciousness. The breath and perspiration also contain water vapor and minute quantities of other contaminants that must be removed by the ECS. Finally, the ECS must be light, simple, and fire resistant.

The best argument for 100% oxygen is simplicity. A two-gas system is not only heavier, because it operates at higher pressure and requires stronger spacecraft walls, but it also is much more complicated, entailing complex sensing and regulating equipment to keep the proportion of each gas constant. And, of course, tanks and other plumbing must be provided for two gases instead of one.

The problem with 100% oxygen in a spacecraft occurs, oddly enough, not in space, but on the ground. The thin walls of a spacecraft are designed to contain gas at a higher pressure inside than out, so that on the ground, with 15 psi outside, the pressure inside is no less, and the spacecraft becomes a tinder box. Pure oxygen at 15 psi contains five times as many oxygen molecules as ordinary air, and will react to sparks and combustibles by an incandescent explosion. NASA took nearly 10 years to verify this, killing three Apollo astronauts in the process, but at the beginning of Project Mercury the problem had not been thought through clearly. The cabin atmosphere had been designed for space, not for the launch pad.

To absorb the exhaled carbon dioxide, the cabin gases passed through a filter containing lithium hydroxide, which combined with carbon dioxide to form lithium carbonate with water as a by-product, or as a chemist would put it: $2LiOH + CO_2 \rightarrow Li_2CO_3 + H_2O$. This part of the ECS was quite simple and well suited for short-duration flights, where the weight of the chemicals does not become excessive.

Mercury's designers did not feel safe with just the ECS. If a titanium weld cracked, or a meteoroid punctured the pressure hull, the astronaut needed backup protection, i.e., a pressure suit. With a ruptured cabin exposed to the vacuum of space, he could still maintain life-saving pressure inside the suit long enough for him to return safely to earth. The pressure suit, with its own oxygen supply and associated hardware, was a miniature ECS in itself, and it functioned like the main system, except at a slightly reduced pressure in order to give the astronaut as much mobility as possible.

Vital as the ECS was, it was a passive system as far as the astronaut was concerned. He made sure it was turned on and working properly, then he left it alone and, if all went well, he paid no more attention to it than he might to the thermostat in his house. The stabilization control system, on the other hand, was more to a pilot's liking. Its function was to control the direction in which the capsule was pointing, and manipulating it was as close to flying the machine as a Mercury astronaut could get.

In an airplane the pilot controls pitch, roll, and yaw by use of elevator, ailerons, and rudder. The Mercury could not use such aerodynamic surfaces in the vacuum of space, but instead controlled its direction, or attitude, by firing tiny hydrogen peroxide rocket motors mounted around the capsule's circumference and pointed in various directions. When the astronaut wanted to raise the nose of the Mercury he pulled back with his right hand on a control stick, and an electronic box selected the appropriate hydrogen peroxide jets to fire. Likewise, by moving his hand to the left or right, he could control roll movement. Unlike an airplane, the Mercury had no rudder pedals. Their yaw function was built into the third axis of the control stick, or "hand controller" as it was sometimes called, and the astronaut had to learn a new movement: rotating his wrist left or right to control yaw, instead of pushing with his feet on rudder pedals.

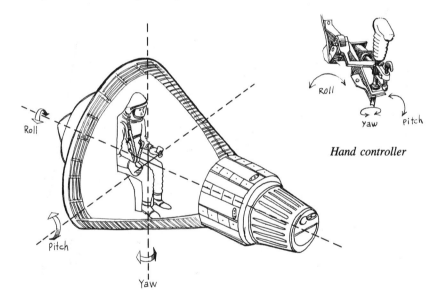

Hand controller

Because of the unknown hazards of space, the Mercury was designed to be flown automatically, so that the capsule could make a successful trip with an empty cockpit, a chimpanzee, or a disabled human. The key components for its automatic functioning were gyroscopes that measured pitch, roll, and yaw angles, and horizon scanners that, as their name implies, were able to sense the dividing line between earth and sky by using infrared heat detectors.

Left to its own devices, the automatic stabilization control system took over as soon as the Mercury separated from the Atlas. It fired jets in a posigrade direction (down the line of the Atlas' flight path) to accelerate and clear away from the Atlas, and then yawed the capsule 180° to the left and pitched the nose down 34°. These maneuvers placed the heatshield forward and slightly

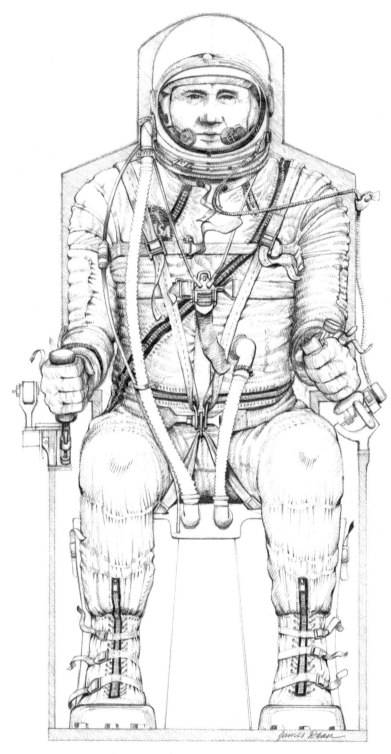

Astro-mummy

above the horizon, in proper position for the retrorockets to fire if an immediate abort from orbit were required. If the astronaut did not like flying around the world backwards, he could take over control, either with electronic assistance ("fly-by-wire") or without it ("direct").

The astronaut sat or lay on a couch custom contoured to his body. The couch, which Carter Collins had ridden in a prototype version to 20 Gs on the Johnsville centrifuge, was constructed of a crushable honeycomb material bonded to a fiberglass shell. The astronaut was firmly restrained in it by shoulder and chest straps, a crotch strap, a lap belt, and toe guards. His arms and hands were held in place by his gripping an abort handle on the left and the hand controller on the right. His knees were bent and elevated slightly. In short, he was positioned to expect the worst, not the free float of orbit, but the jarring deceleration of a steep abort trajectory.

Encased in a bulky pressure suit, hog-tied in his seat, with bulkheads within inches in all directions, the astronaut was bundled up like a mummy wrapped inside a tight sarcophagus. Directly in front of him was the instrument panel, from which peered the lens of a movie camera that recorded his every movement, even the tic of an eye. Aeromedical instrumentation consisted of monitor circuits for electrocardiograph, respiration signals, body temperature, and blood pressure. Blood pressure was measured using an inflatable arm cuff, while respiration rate and depth were derived from a helmet-mounted thermistor. The electrocardiograph signals came from four sensors taped to the astronaut's chest, and—a final indignity—body temperature was taken using a rectal probe. The information from all these gadgets was telemetered to flight surgeons on the ground.

The instrument panel was airplane-like in its general configuration, and in fact many of the Mercury instruments were unmodified airplane gages. Despite the weight battle, miniaturized custom-built switches, controls, and instruments were rarely used. John Glenn refers to the large off-the-shelf instruments as "old steam gauges," which I suppose is a battleship term. It does seem a bit incongruous to have a tiny capsule fitted out with huge knobs and dials. Prominently displayed were pitch, roll, and yaw angles, G meter, altimeter, rate of descent indicator, cabin pressure gauge, voltmeter, ammeter, and a fancy clock—all pretty standard stuff.

Much less conventional was an 8-inch periscope assembly through which the astronaut could observe the earth through the floor of the capsule. It had two settings, low and high magnification. The astronaut had a field of view of 1,900 miles on low and 80 miles on high. He could use yellow or red filters, and could rotate a sun-moon index to help orient himself. He could also bypass the periscope and peer out through either of two tiny portholes.

Another unusual instrument was the earth path indicator, which consisted of a 4-inch spherical globe representing the earth. It rotated in such a manner that the astronaut was presented with his position over the earth. The McDonnell

Corporation duly noted that the following features had been included: "all continents, all bodies of water having major dimensions of 300 statute miles, the 16 largest rivers in the world, all islands having major dimensions of 500 statute miles, . . . the 50 largest cities of the world identified by .020-inch dots. The touchdown area is displayed as a rectangle and the luminous dot inside the rectangle is the point of impact. The landing area is 3,040 nautical miles ahead of . . . orbital position . . . indicated by the four-ring bull's-eye."

The instrument panel also contained various switches and controls for monitoring and regulating the electrical system, which was battery powered, and several radios, as well as the telemetry equipment that relayed the condition of astronaut and critical hardware components to ground stations.

In an emergency, the astronaut could use his left hand to rotate the abort handle and blast the capsule free of the Atlas by firing the escape tower rocket. This tower was a 10-foot-high steel frame supporting a single solid rocket motor that fired through three separate exhaust nozzles canted slightly outward. On the launch pad, its 56,000-pound thrust could yank the Mercury capsule up to an altitude of nearly half a mile, sufficient for a safe parachute deployment and descent. The original Mercury concept did not include such a device, but the Atlas' woeful safety record soon caused Faget to add it. Gilruth and his Langley people had used solid rockets for a long time and trusted them implicitly, unlike their feelings toward the temperamental Atlas, which pumped huge quantities of liquid oxygen at minus 297°F.

Mercury's descent was controlled by one 63-foot diameter main parachute backed up by a reserve. There was nothing new about this technology, but the development and procurement of these parachutes nearly drove Caldwell Johnson crazy. "The parachute industry was comprised of old guys doing seat-of-the-pants calculations and old ladies with sewing machines. No two parachutes turned out the same. You had to tell them 'When you run out of green thread, don't replace it with a different color.' " On Alan Shepard's flight Gilruth insisted that Shepard add a personal chute to an already crowded cockpit, as last-ditch insurance against failure of the capsule parachutes.

Recovery of the Mercury was planned in the Atlantic Ocean southeast of Bermuda. Because of its launch azimuth and its point of origin in northern Florida, its trajectory could not reach any farther north than 28° in the Northern Hemisphere or farther south than 28° in the Southern. However, considering all the abort possibilities, it was conceivable that it might come down in a jungle or desert or anywhere else on the earth's belly band over which it crisscrossed. If that became the case, it might be many hours before help arrived, and the astronaut was provided with a survival kit to tide him over.

Stowed just to the left of the couch, the kit contained two dozen items, including: a life raft, shark repellent, food and water, matches, pocketknife, whistle, signal light, rescue radio beacon, a bar of soap, and three antimotion sickness injectors. The idea behind the injectors was that the astronaut might become so violently ill in space or on the water that he would not be able to swallow a pill and keep it down. Instead, the spring-loaded injector was designed to penetrate the tough layers of his pressure suit and plunge its needle deep into his thigh. Unwilling to test it on himself but curious about its efficacy, John Glenn took one into his crew quarters one night at Cape Canaveral and gingerly tried stabbing a grapefruit. It worked perfectly, but he wasn't sure whether that was reassuring or not.

Although he knew it would be crowded inside his tiny cockpit, John Glenn thought, like any good tourist, that he should take along a camera. His suggestion was greeted with objections from all sides. "It seems crazy now," Glenn told me, "but people were worried about any extra weight, and they thought I shouldn't be distracted from my other tasks in any way. I went to Gilruth, and he agreed it was ridiculous not to get any pictures, so I asked a photographer friend of mine to recommend something." Based on this, NASA procured a 35mm Minolta for $45 from a Cocoa Beach drugstore, and it was officially added to the equipment list.

Of all the Mercury Atlas components, the most intriguing subsystem was the astronaut himself. "The capsule crew consists of one man representing the peak of physical and mental acuity, training, and mission indoctrination. Much more will be required of the crewman than is normally required of the modern aircraft test pilot. The crewman must not only observe, control, and comment upon the capsule system, but must scientifically observe and comment upon his own reaction while in a new, strange environment." This was the somewhat gushy view that McDonnell published in a training manual.

However, before formal selection criteria were established in January 1959, the situation was not nearly so tidy, especially regarding the comparison with test pilots. No one was quite sure what an astronaut could or should be. Gilruth was flexible. After all, had he not gotten a weight lifter to fly his P-47? Nonetheless, he was somewhat taken aback at some of the suggestions he received, such as the one from industry that a double amputee be selected to save weight.

Macho seekers of danger, such as bullfighters and mountain climbers, were also frequently suggested. The Space Task Group called upon its aeromedical consultants, and they proposed a wide variety of professions that might be suitable. Three characteristics were required: "(a) willingness to accept hazards comparable to those encountered in modern research airplane flight; (b) capacity to tolerate rigorous and severe environmental conditions; and (c) ability to react adequately under conditions of stress or emergency." Scientists, technicians, aviators, submariners, divers, parachutists, and mountain climbers were all to be considered, provided they had proper credentials in terms of education, work experience, and sex (males only). Naturally the aeromedical consultants did not ignore medical doctors, nor test subjects who had participated in medical or environmental investigations.

Whatever their backgrounds, the paragons selected were to be paid between $8,330 and $12,770 per annum, "depending on qualifications."

The donnybrook shaping up between disparate disciplines and their support groups was abruptly ended by President Eisenhower. Over the 1958 Christmas holidays he decided that military test pilots formed a sufficiently large pool of preselected talent from which to draw. Test pilots, because they had already volunteered for and proven themselves in a high risk profession; military, because Mercury might involve classified work and military officers already possessed the necessary clearances. Seven criteria were announced to the public:

1. Age - less than 40
2. Height - less than 5'11"
3. Excellent physical condition
4. Bachelor's degree or equivalent
5. Graduate of test pilot school
6. 1500 hours total flying time
7. Qualified jet pilot

There should be a large asterisk beside number 5, because that is what defined the competition. The Navy at Patuxent River, Maryland, and the Air Force at Edwards, California, had the only test pilot schools in this country, and each produced only a couple of dozen graduates a year. The Pentagon reviewed 500 records and estimated that it had 110 qualified candidates.

What were these men to do? NASA's literature said, "Although the entire satellite operation will be possible, in the early phases, without the presence of man, the astronaut will play an important role. . . ." Well, if it could be done *without the presence of man*, how important could his role possibly be in the operation of Mercury? Even more degrading to the egotistical test pilots was the notion that chimpanzees would precede them. Years later Chuck Yeager recalled, "I wouldn't want to have to sweep off monkey shit before I sat down in that capsule." Other test pilots put the astronaut's role in only slightly more delicate terms: "Spam in a can."

On the other hand, many test pilots were intrigued with the notion of flying higher and faster, irrespective of the fact that, on Mercury at least, their control over their destiny would be minimal. John Glenn, for example, genuinely felt then and feels now that flying in space is a clear promotion for any pilot. Prior to his three orbits in Mercury, Glenn had set an intercontinental speed record in a Crusader jet fighter. I asked him, as an aviator, which machine he preferred. He definitely ranked the primitive Mercury above the sleek supersonic perfection of the Crusader because, he said, the Mercury flight was of an entirely different nature, a new realm. Wings and wheels were immaterial, it was getting up into space that was important. Mercury carried with it international prestige that the other did not, he went on. It was important not only to beat the Russians, but to establish long-term research objectives in space.

No poll was taken among test pilots, but a clear majority of those eligible for Mercury did volunteer. Of these, 32 advanced to the stage of taking physical exams, and the seven winners were announced to the world at a press conference on April 9, 1959. There were three from the Air Force, three Navy and one Marine. Air Force Captains Donald K. (Deke) Slayton and L. Gordon Cooper came from Edwards, the mecca of Air Force flight testing, while Captain Virgil I. (Gus) Grissom had been testing all-weather fighters at Wright-Patterson Air Force Base, Ohio. Navy Lieutenant Commanders Alan B. Shepard, Jr. and Walter M. Schirra, Jr. had put in tours at Patuxent River, as had the senior member of the group, Marine Lieutenant Colonel John H. Glenn, Jr. Navy Lieutenant Malcolm Scott Carpenter's background was slightly different from the others in that he had specialized in multiengine transport planes rather than jet fighters.

Probably no other group of men has been so prodded, probed, pierced, and psyched. During a week of exhaustive medical examinations, every conceivable measure of their health and physical performance was chronicled. Then followed another week of studies and measurements under various conditions of real and imagined stress. They were baked, chilled, vibrated, isolated, pressurized, and quizzed about their innermost thoughts. "Nothing is sacred anymore," one complained. They were evaluated by a board of eight, chaired by an engineer, and which included a test pilot, two flight surgeons, two psychologists, and two psychiatrists. The board weighed not only their physical and psychological strengths, but also their technical training and flight experience. As test pilots, they ranged from experienced (Slayton) to novice (Cooper), but as physical and mental specimens they were the cream of an already highly select group.

The "Original Seven" as they came to be called after other groups of astronauts were picked, moved to Langley and went to work. In addition to their basic indoctrination, each was assigned a technical area in which to specialize, so that collectively they could scrutinize the entire Mercury setup—plans, hardware, schedules, safety margins, crew's role, etc.

They swiftly became the darlings of the press, who seemed unable to satisfy the public appetite for these new celebrities about to zoom off into some exotic unknown. Dryden and Gilruth tried unsuccessfully to moderate this deification process, but the Seven's influence grew far out of proportion to that of ordinary test pilots planning the first flight of a new plane. For example, within 2 months of their selection they were testifying before a Congressional Committee, passing judgment on the status, progress, and safety of Project Mercury, subjects about which they knew woefully little at that time.

They also visited contractors' factories, where their opinions were gravely received. John Yardley recalls that "on their early visits to McDonnell they were not the Gods they became later, but their inputs were considered." Their "inputs" to McDonnell included a request for a large window in the side hatch, instead of submariner Faget's two minuscule portholes, and a way of explosively removing the side hatch to make egress quicker and safer than through the skintight passageway past the instrument panel to the apex of the capsule. They reviewed in detail the function and arrangement of the 120 controls, 55 electrical switches, 30 fuses, and 35 mechanical levers in the cockpit, and recommended a number of changes. Their requests were approved quickly. There was also a strong public relations flavor to their factory visits, the theory being that a worker who actually chatted with one of them would be extra careful not to kill the fellow by doing slipshod work.

Aside from the factory visits, they spent long hours preparing themselves for flight. They flew jet trainers whenever they could, mostly on weekends, and like most pilots they groused about the shortage of flying time. They carried a heavy training load in the form of classroom lectures and simulator exercises, during which they not only learned all the subsystems of the capsule but also worked out normal and emergency procedures and checklists.

The managers of the Space Task Group, accustomed to aircraft research programs, genuinely believed that the astronauts were not just guinea pigs, but active and valuable participants in the safe operation of the machine. Or as Scott Crossfield, one of the country's most experienced test pilots, put it: "Where else would you get a nonlinear computer weighing only 160 pounds, having a billion binary decision elements, that can be mass produced by unskilled labor?"

Nonetheless, the first Mercury crew member was not a 160-pound human being, but a 37-pound, 4-year-old chimpanzee named Ham. First using replicas of the Mercury capsule called "boilerplates," and then unmanned versions of the Mercury itself, NASA carried out increasingly sophisticated tests during 1959 and 1960 using the Little Joe and Redstone rockets. By early 1961, the astronauts felt fully trained and ready to fly, but NASA was not about to risk one of the group whose names had become household words, without convincing evidence that they could survive the space environment.

Some medics were still unsure about man's value to the Mercury flights, and argued that the astronaut should be anesthetized or tranquilized so that he wouldn't interfere with any of the vehicle's automatic functions. Ham's suborbital lob aboard a Mercury Redstone dispelled some of this doubt. Weightless for nearly 7 minutes, Ham not only survived in good health, but telemetry indicated he had performed all assigned tasks well even during his 14.7 G entry. The chimpanzee had been trained to respond to light signals by pulling a lever. If successful he was rewarded by banana pellets. If he failed, he received an electric shock through the soles of his feet. Much to the chagrin of the astronauts, their old test pilot buddies made many bad jokes about what astronaut duties might entail.

Within 3 months, Ham was followed by a human, but his name was Gagarin—Major Yuri Alekseyevich Gagarin, age 27, a Soviet Air Force pilot and parachutist. His Vostok 1 ("East") capsule was three times as heavy as the Mercury, and Gagarin had made one complete orbit of the earth, not just a suborbital lob.

It was devastating news to the Space Task Group and to the country. Twenty-eight years later John Yardley still remembers: "Yuri Gagarin pissed me off." Lieutenant Colonel John A. "Shorty" Powers, spokesman for the astronauts, blurted out what the public was rapidly concluding: "We're all asleep. . . ."

Actually the news was not a big surprise to Gilruth. "We knew the Russians were doing the same as we were, but we didn't have any intelligence about their progress. But I knew, because when I met some of them at an international conference they'd ask: 'What vertical velocity at landing are you designing for?' "

The fundamental question was to what extent Gagarin's flight expressed the overall progress of science and technology in the Soviet Union. As early as 6 days after Sputnik I, Walter Lippmann had written: "Their being so much ahead cannot be the result of some kind of lucky guess in inventing a gadget. It must be that there is a large body of Soviet scientists, engineers, and production men, plus many highly developed subsidiary industries, all successfully directed and coordinated, and bountifully financed." In the 3½ years between Sputnik and Vostok 1, it appeared that the Russians had at least maintained, and probably increased, their lead. "The Russian accomplishment was a great one," Glenn acknowledged.

The Space Task Group made some minor changes as a result of Vostok 1, eliminating or accelerating some tests, but basically they felt that Mercury was a well-conceived program and they stuck to their guns. They would not go directly to a manned orbital flight aboard the Atlas, but would first test man—like Ham—aboard the Redstone in a suborbital trajectory.

Alan Shepard had been picked for the first flight, with Glenn and Gus Grissom his back-ups. Although the three of them knew Shepard was the one, the public was told that all three were contenders, creating a split of the seven into the Inner Three and the Outer Four, with the press clamoring for the former and ignoring the latter. Naturally none of them was pleased with this arrangement except Shepard. Al had been picked by Gilruth and company based on his performance in the simulators and their general impression of his competence. Yardley recalls simply that, "Shepard was the sharpest." His selection, when it was finally made public, subjected him to even greater scrutiny but, a couple of years after this, I remember Al's great jawbuster of a grin: "I still have a few secrets."

Shepard and his Mercury Redstone were ready in less than a month after Gagarin. His flight was designated MR-3, MR-2 having been Ham's, and MR-1 an ignominious flop during which the Redstone had moved all of 6 inches off the pad. "The day Shepard was launched I was in the blockhouse. All I could think of was that . . . Redstone . . . 6 inches," John Yardley recalls. But at 9:30 A.M. on May 5, 1961, the Redstone worked perfectly and catapulted Shepard to a speed of 5,000 mph and an altitude of over 100 miles. He landed 300 miles downrange from Cape Canaveral after a 15-minute flight.

Because of prelaunch delays, due to weather and mechanical problems, Al had had to wait inside his capsule, which he had named Freedom 7, for 4 hours. Finally his bladder rebelled and he became not only the first American but, with no report to the contrary from Gagarin, the first "wetback" to venture into space. Aside from this minor indignity and some flight mismanagement of the attitude control system on his part, Al and his equipment performed brilliantly.

Like Gagarin, he seemed to have suffered no ill effects, and the most dire predictions of the medics were put to rest. He swallowed perfectly, he thought clearly; his heart had not stopped, nor his vision dimmed. He enjoyed seeing large chunks of Florida and the Bahamas through his periscope but he couldn't see much through the two tiny portholes of his unmodified capsule. Hereafter, all flights included the large side hatch window proposed by the astronauts.

Shepard's flight was routine in the sense that his equipment worked properly and there were no significant deviations from preflight predictions, but it certainly was anything but routine in comparison with the flight of an airplane—even a high performance experimental rocket plane.

First, consider the rapid acceleration, from zero to over 5000 mph in less than 3 minutes. Next the transition through the region known as "max-Q," or maximum dynamic pressure. In this zone the ram impact of the air reaches its greatest intensity and, coupled with compressibility effects of approaching Mach 1, causes the spacecraft to shake and rattle. Even though tightly restrained, Shepard's head was bounced around so severely that for a short time he was unable to read his instruments. Follow this with 5 minutes of weightlessness, far more than an aircraft can experience, and with an unparalleled view. Then fire the retrorockets and dive back into the atmosphere with a deceleration of nearly 12 Gs, followed by parachute deployment and a resounding splat into the ocean.

A lot happened during Shepard's flight, and though he didn't control it all the way, as his test pilot compadres did with their jet airplanes, he experienced a wilder ride in 15 minutes than most of them would in a career. And when he did fly the capsule manually, in the vacuum of space it was possible to turn in any direction, using whatever combination of roll, pitch, and yaw angles he chose, unlike an airplane which basically must plow through the air nose first. With a test pilot's typical understatement, to Al Shepard all this was "just a pleasant ride."

Gus Grissom's turn was next, 2 months later. As with Al, he had the choice of naming his capsule, and picked Liberty Bell 7. In a display of astronaut group unity, the number 7 followed all the Mercury names. Grissom's capsule included some modifications, chief among them an explosive side hatch with a large window in it, but basically his trip was a repeat of Shepard's. Grissom was allocated more time for piloting duties, maneuvering the spacecraft and observing out the new window, and he was mightily impressed by the totally black sky and 800-mile arc of horizon visible from 100 miles up. Gus's pulse shot up to 162 beats per minutes before launch and 171 beats during entry.

Once on the water, when it should have been all over, Gus' problems began. The button that detonated the explosive side hatch was protected in flight by a cover over it and a safety pin through it. With the rescue helicopter nearby but not directly overhead, Gus prepared for his exit by removing the cover and the pin from the detonator. He did not remember pushing the button. In fact, as time went by, he became more adamant in the belief that he had not. "I didn't touch it. I was just lying there—and it blew." With the side hatch gone, heavy swells began filling Liberty Bell 7 with water.

Fortunately, Gus had already disconnected his suit from the spacecraft, and he was able to exit through the hatch quickly and swim away from the sinking capsule. Unfortunately, he had forgotten to close an oxygen inlet valve, and his heavy but normally buoyant suit began filling up with water. He, too, was sinking and he began waving for the helicopter. The helicopter pilot, however, figured Gus was signaling that he was all right and devoted his attention to snagging Liberty Bell 7. Full of sea water now, the spacecraft was too heavy to lift and the helicopter eventually lost its tug of war and cut it loose. By this time a second chopper had picked up an exhausted Grissom. The Air Force captain had lost his first naval command.

The matter of the hatch jettison was never satisfactorily explained. Ground test failed to duplicate the sequence, and with Liberty Bell 7 three miles down in the Atlantic, the actual hardware could not be examined. The problem was solved simply by a change in procedure: henceforth, the cover and pin would remain in place until the helicopter had a firm grip on the capsule. Just as pilot Grissom was sure he hadn't touched the jettison button, so was engineer Yardley convinced Gus must have bumped it, an example of the tension, perhaps inevitable, that exists between designers and operators of complicated machines.

Despite the loss of Liberty Bell 7, NASA decided it had learned enough from the Redstone lobs and was ready to undertake orbital flight. Just to rub it in, before NASA could make its next move, the Russians sent up Gherman Titov in Vostok 2. Titov leapfrogged from Gagarin's one orbit to *17*, staying up for 25 hours. By comparison Project Mercury seemed puny indeed. Nor was NASA going to rush into the more complex problems of a Mercury Atlas flight just because the Russians were ahead. Before committing a man, NASA insisted on another ape flight, despite a fair amount of criticism, in order to validate Mercury's ability to handle the higher speeds and temperatures of an Atlas orbital flight.

Enos was a prince of an ape, a native of the French Cameroons, who in a tough competition had beat out the veteran Ham and a half dozen others in a process not unlike the one Al Shepard had undergone. Like Shepard, Enos had endured long sessions on the centrifuge and the procedures trainer, and had learned to perform a complex series of tasks. He would be rewarded not by phone calls from the President but by a drink of water and a banana pellet.

Enos was launched on November 29, 1961, into a near-perfect orbit and commenced circling the earth once every hour and a half. Instead of looking out the window, he was kept busy pulling levers in response to various cues programmed by the medics. In the midst of these tests, however, one of the levers malfunctioned and the poor little bastard received nearly 80 undeserved shocks. Because of minor problems with the ECS and the attitude control system, Enos was brought down after two orbits instead of the planned three. Nonetheless, the flight was deemed a success and Enos, good as new, was sent off into retirement to contemplate the perfidy of men who, promising banana pellets, delivered instead electrical zaps through the soles of one's feet, even though he had played their silly game to a fare-thee-well. He should have been awarded the NASA Exceptional Service Medal.

NASA was finally ready. Ham, Shepard, Grissom, Enos—all were preliminaries to the main event, putting a man in orbit. Had it not been for the damned Russians, with their Gagarin and Titov, it would have been a moment of great national triumph. As it was, it was an opportunity to salvage somewhat NASA's and the country's reputation for technological leadership, and in the full glare of international publicity, not skulking off from some remote launch base in Kazakhstan, out of sight of all but a few workers and party apparatchiks.

John Glenn was certainly ready. After months in the almost servile role of back-up to Shepard and Grissom, he was chomping at the bit. He had studied

Enos ready for launch

the flight plan, memorized emergency procedures, flown the simulator. It seemed as if he had attended a thousand meetings and nearly as many press conferences. He wasn't quite sure what to make of this role. Using a test pilot's perspective, he had already lost the game to Shepard: after all, the first flight of a new machine was it! But was it, in this case, when the first two flights hadn't really gotten there? Maybe Shepard and Grissom had each made a taxi test and he was making the first flight.

The press seemed to think so, implying that Glenn had been saved by the authorities for the main event. "The Marine" had been their favorite from the outset. He exuded all the middle American virtues we look for in our heroes, even down to the freckles. Compared to the acerbic Shepard and the taciturn Grissom, he seemed so genuine, so wholesome, so all-American. It didn't embarrass him to be patriotic. On the contrary, unlike the other six, he seemed to enjoy talking about himself and about Mercury in the context of a nation's pride in its history and its future achievements.

The machinery also seemed ready to accompany Glenn. The capsule itself had come through its various preliminary tests in good shape. A few leaks, a few stuck valves, some minor cracks, but the basic design—Faget's shape, the escape rocket, the ablative heatshield, the retrorockets, the parachutes—all had performed as well or better than expected. The Atlas was less certain. Its overall safety record was certainly not an enviable one, but it was getting better, and the last two flights of the man-rated version had been successful. Besides, if the Atlas failed, there was always the escape tower with its faithful solid rocket motor. The machinery was as ready as it was going to get.

Unfortunately the weather at Cape Canaveral was beyond NASA's control. Despite what the Florida Chamber of Commerce says, it can be terrible in the winter, and January 1962 was worse than most. Time after time, Glenn's flight was delayed, once by an Atlas fuel leak, and then again and again by overcast skies. The 600 accredited news people groaned at the delays. They seemed more exasperated than Glenn, who explained patiently that delays were inherent in any flight test program.

Finally, on February 20, the weather cleared and Glenn was off. An estimated 100 million people watched on television as the Atlas, streaming a long tail of fire, arced out over the Atlantic and streaked off to the east. Friendship 7's ride was smooth at first, got "a little bumpy out here" at max-Q, then smoothed out again. When the escape tower was jettisoned Glenn got his first look out, "a beautiful sight looking eastward across the Atlantic." He was in a good solid orbit, apogee 160 miles, perigee a safe 100. The Atlas had come through.

For three orbits, each slightly less than an hour and a half long, Glenn went about his tasks in a serious, businesslike manner. He liked weightlessness but at the same time it frustrated him to be so locked in by form-fit couch, pressure suit, and straps that he couldn't fully savor the sensation. He felt none of the queasiness that the Russian Gherman Titov had reported. And he loved the view, seeing all of Florida in one glance, just like on the page of an atlas. He flew facing backwards, forwards, and sideways, troubleshooting the attitude control system, which had a malfunctioning thruster. But this problem wasn't serious, and by analyzing it he felt more in charge of things, more like a test pilot. Likewise some minor overheating of the ECS required his attention. He became the third human to experience a space sunrise and sunset—each less than an hour apart. "That sure was a short day," he told Gordon Cooper as he passed over Australia.

The thing that intrigued Glenn the most was what he called the fireflies. "I am in a big mass of some very small particles that are brilliantly lit up like they're luminescent . . . they look like little stars . . . they swirl around the capsule." While Glenn was preoccupied with these strange visitors, his compatriots on the ground were wrestling with quite a different problem. One of their telemetry signals indicated that the heatshield and landing bag had somehow come loose and were attached to the capsule only by the straps holding the retrorockets in place. Was it faulty telemetry, or was the heatshield really dangling? If it was, the consequences could be catastrophic during entry—capsule and Glenn incinerated, a blazing meteor.

The experts huddled in Mission Control to consider their first crisis. Assuming that the heatshield really was loose, what could be done? Max Faget reported by telephone that the retropack, consisting of three solid retrorockets, could be retained after firing instead of using the normal jettison procedure. The straps attaching the retropack to the capsule should hold the heatshield more or less in place until the straps burned up during entry. By that time, air pressure on the heatshield should be sufficient to keep it plastered up against the capsule.

The only hitch, Faget warned, was that if only two retrorockets fired upon command, the third would spontaneously ignite during entry, with unknown, but possibly catastrophic, results. As a test, Glenn was asked to manipulate the landing bag switch, and the results indicated that the problem was probably a false indicator, but still, no one could be sure, and NASA decided on a conservative approach, and told Glenn to retain the pack instead of jettisoning it after retrofire.

No one spelled out for John exactly what the problem was, but what with all the fussing over the landing bag switch, plus a couple of people asking him whether he felt anything banging up against the side of the capsule, he had a pretty good idea of what was going on.

When retrofire time came, the first hurdle was passed: all three rockets fired. Nonetheless, things stayed exciting for a while. No simulator had prepared Glenn's eyes and ears for what was happening now. He heard "small things brushing against the capsule." He saw "a real fireball outside." A strap from the retropack flashed over his window and disintegrated in a streak of smoke. John thought he saw chunks of retropack flying by and feared that the heatshield itself might be breaking up. Friendship 7 began to roll and yaw, bucking like a panicked pony. Glenn couldn't damp this motion manually, so he called on the automatic control system for help. The oscillations quieted down then, but he was burning up attitude control fuel so fast that both the automatic and manual systems were headed toward empty.

First the one, then the other, went dry and still he had a minute to go before the drogue parachute was scheduled to deploy. With neither drogue nor fuel for stabilization, might not the capsule flip over backwards and foul the parachutes as they came out? Just then the drogue deployed, the oscillations stopped, and John realized he had survived the entry. The main parachute popped on schedule, and Friendship 7 splashed comfortably into the Atlantic.

Friendship 7's entry fireball

None the worse for wear after his wild ride, Glenn was pronounced fit by the medics and congratulated by President Kennedy.

Glenn was somewhat miffed by the fact that his compatriots on the ground had not shared fully with him the details of the false alarm about the landing bag and heatshield. The more informed the astronaut, he felt, the better he could cope with any inflight emergency. An ''include the test pilot'' rather than a ''don't scare the passenger'' philosophy evolved from his incident and has served NASA well over the years of Gemini, Apollo, and Space Shuttle.

From a mechanical, as well as a human, point of view, Glenn's flight was a resounding and reassuring success. Considering that each Atlas and each Mercury contained around 40,000 critical parts, both had performed brilliantly. Glenn's capsule had grown from 2,700 pounds to 3,000 pounds, but the Atlas' performance had kept pace, and at liftoff the total Mercury Atlas system weighed 260,000 pounds and produced a thrust of 360,000 pounds. The Soviet equipment might be heavier and more powerful, but President Kennedy, in a Rose Garden ceremony for Glenn, made it very clear that the competition had just begun: ''We have a long way to go in this space race. But this is the new ocean, and I believe the United States must sail on it and be in a position second to none.'' And the nation applauded. Four million people turned out in New York City for a parade on ''John Glenn Day.'' A joint session of Congress listened enthralled to his debriefing.

Although Glenn's three orbits must be considered the apogee of the Mercury program, the three flights which followed it were very important in establishing a solid institutional base for NASA. It was not to be a flash-in-the-pan agency, it was embarked on a long-range course to the moon, and beyond, and it needed to build a system for regularly and successfully operating in space. Four of the Original Seven astronauts had yet to fly. One of them, Deke Slayton, had been disqualified by a minor heart ailment, but the other three were salivating for their turn, not to mention Shepard and Grissom, who wanted seconds.

Scott Carpenter, who had been Glenn's back-up, was next, in May 1962, aboard Aurora 7. Reflecting NASA's conservatism, his flight was essentially a repeat of Glenn's three orbits, but with the addition of a number of experiments, such as towing a balloon to obtain aerodynamic drag data near perigee, and photographing the horizon using red and blue filters to define the horizon line more precisely for future navigators. Carpenter's sense of wonder and delight over spaceflight seemed to exceed that of the other astronauts. He was so taken with his new playground that at times he had to be reminded to perform his routine duties, and he badly mismanaged the attitude control system, wasting fuel. But the worst thing he did was to fire the retrorockets while misaligned in yaw by a horrendous 25°. He also fired them 3 seconds late. These two errors, coupled with something beyond his control, slightly underpowered rocket motors, combined to cause him to overshoot his landing point by 250 miles.

Like Glenn, Carpenter ran out of attitude control fuel during entry, and had a rough ride because of it. Unlike Glenn, he didn't have chunks of flaming retropack to worry him. In fact, he rather enjoyed the whole experience and when the frogmen eventually reached him, they found him comfortably ensconced in his raft, offering food and drink to the visitors. Flown back to Grand Turk Island, he talked far into the night, explaining such things as the fireflies, which he had discovered were ice flakes on the skin of the capsule that could be dislodged by thumping on the hatch.

Wally Schirra was a lot different. His philosophy, frequently stated, was that one had to "pull with an even strain." By that Wally generally meant that he wasn't going to overexert or overextend himself. In this case, it meant that, badly as he wanted to fly in space, he knew he had some leverage over the system and he was going to use it to make sure a bunch of fruitcake experiments weren't going to mess up his main reason for going, which was to make a dignified, leisurely, and flawless test flight. Where Carpenter, and to a lesser extent Glenn, had been rushed to the point of frenzy, Schirra's flight plan by comparison was a desert of rest and contemplation interspersed with placid oases of controlled activity—piloting activity.

Mercury's objectives had been met by Glenn: to make sure that a man could survive safely and operate effectively in space. Beyond that, it was all icing on the cake. Sure, we couldn't forget about the Soviets, but let's not get frantic about it, let's . . . pull with an even strain! Besides, the Soviets seemed to be operating in a different realm. Only 2 months before they had put up not one but two Vostoks within 24 hours, and they had orbited within 3 miles of each other. Even worse, one of the cosmonauts, Andrian Nikolayev, had stayed up for 4 days! We couldn't come close to that, so we might as well pull with an even strain for just 6 orbits, which was all the Mercury managers were willing to attempt without modifications to several of the capsule's systems. Among other challenges, Wally would have to ration his attitude control fuel carefully to avoid running out, as Glenn and Carpenter had done.

Considering his flight to be primarily an engineering and piloting evaluation, Schirra emphasized this point by naming his capsule Sigma, the Greek symbol used in engineering parlance to indicate summation, or totality. Of course he added 7 to the name, as the others had done. Sigma 7 was launched on October 3, 1962, and went smoothly from the very beginning. Wally used only 1.4 pounds of attitude control fuel in his first orbit, a remarkably small amount which can be attributed to his piloting skill and the fact that his flight plan didn't call for any strenuous maneuvering. After determining that he could measure his all-important yaw angle within a couple of degrees, day or night, simply by careful observation out the hatch window, Wally concluded that the heavy periscope was unnecessary, and it was not flown again.

Wally spent a lot of time in what he called "chimp mode," letting the automatic system fly the capsule, and for even longer periods he turned the attitude control system off, and allowed the capsule to drift aimlessly, turning slow cartwheels in the sky. This situation reminded Wally of an old song, "Drifting and Dreaming." He also exercised a bit, pulling on a bungee cord, and he spent a lot of time fiddling with the ECS, trying to get the suit circuit to stabilize at the most comfortable temperature.

Just prior to retrofire, he still had 78% fuel remaining in both the automatic and manual attitude control systems. Instead of Carpenter's miss distance of 250 miles, Schirra landed within 5 miles of his target. Carpenter never flew in space again. Schirra flew on Gemini, and when Gus Grissom was killed, he took his place and made the first test flight of the Apollo Command Module, becoming the only astronaut to make a Mercury, Gemini, and Apollo flight. Gene Krantz, one of the flight controllers, congratulated Wally while Sigma 7 was still airborne: "Now that's what I call a real engineering test flight!" Wally later agreed: "It was a textbook flight. The flight went just the way I wanted it to."

The final flight of the Mercury series went to Gordon Cooper, at 36 the youngest of the seven. Like Schirra, he had a more relaxed attitude about spaceflight than some of his more intense predecessors, but of course it was precisely because of their success that he could afford to relax. On the other hand, the ICBM version of the Atlas, only slightly different from Cooper's, was still blowing up with some regularity, so what Gordon was undertaking in Faith 7 was certainly no piece of cake.

His objective was to stay up 22 orbits (34 hours), a duration which far exceeded the original Mercury specification, and which required McDonnell to make major modifications to the capsule. Chief among them was the addition of heavier batteries, an extra oxygen bottle, more water, and more hydrogen peroxide fuel. To compensate for these weight increases, the 76-pound periscope was deleted along with some communications equipment. Capsule weight, however, inexorably pushed up past 3,000 pounds despite the engineers' success in rejecting additional scientific equipment. Astronauts and STG engineers alike considered the Mercury finale to be a long duration test flight rather than the orbiting of a scientific laboratory. The capsule weight finally stabilized at 3,026 pounds.

On the launch pad, at a time when Gus Grissom's heart was churning away at 162 beats per minute, Gordo fell asleep. During the second orbit, when John Glenn on his flight was struggling with a malfunctioning thruster, Gordo nodded off again. His workload consisted primarily of conducting experiments and gathering medical data. He launched and photographed a small sphere equipped with a flashing beacon, a test of his visual acuity. He measured the radiation inside the capsule and took his oral temperature and blood pressure. He exercised on schedule and collected urine samples. He photographed parts of South America, Africa, and India. Tibet, in particular, fascinated him, for the air was thin and clear there, and he could make out individual houses and the smoke coming from their chimneys. His photographs were the best taken so far.

After the flight there was much discussion about the unexpected detail of

Gordo's observations, and Gordo encountered some skepticism about his extraordinary eyesight. Somehow out of this came the myth that the Great Wall of China was especially visible, even though Gordo had not flown far enough north to see it, and, a few years later, this notion had grown to the point that it was supposed to be the only man-made feature on earth visible from the moon, which I can attest is certainly not the case.

It was only toward the end of his flight that Gordo had an opportunity to put his test pilot skills to work. Little by little electrical problems began cropping up. First the ".05 G" light came on, a false indication that he was entering the atmosphere, then all attitude readings were lost. Next a short circuit disabled the automatic stabilization and control system. Cooper took all this in stride and remarked only, in laconic test pilot fashion, that "things are beginning to stack up a little." He ended up being forced to make a manual entry, which pleased the pilot in him, and he splashed down even closer to the aircraft carrier than had Wally.

Like Glenn and Schirra, Cooper was slightly dehydrated after his flight, having lost 7 pounds in the day and a half. Also like his predecessors, he suffered from a slight case of orthostatic hypotension, a fancy term to describe the fact that when he stood up, blood tended to pool in his legs, making him feel lightheaded. Other than that he was in great shape, mentally and physically, and ready to fly in space again at the first opportunity.

So were the others. Cooper's back-up, Al Shepard, in particular, felt that one more Mercury flight—his—was in order. However, NASA's new Administrator, James E. Webb, and most of his staff thought that enough was enough. Mercury could be stretched past 34 hours but why risk it, given the fact that the two-man Gemini was coming along, with its more reliable Titan booster. Clearly Project Mercury had met or exceeded all its goals. Only the knowledge that the Soviets had an even greater capacity for spaceflight dampened the participants' enthusiasm.

Project Mercury took slightly less than 5 years and cost $400 million, about half of it the price of flight hardware, and the other half ground facilities. It had lagged considerably behind its original optimistic schedule and it had been described by a Congressman as a "Rube Goldberg contraption placed on top of a plumber's nightmare." Nonetheless, it had come through and passed all its tests with flying colors—and without loss of a pilot, which was more than could be said for other programs to which it was sometimes compared, such as the X-15. Despite some medical silliness along the way (Al Shepard's chest was tattooed with tiny dots to mark the desired location of his cardiac sensors), it was now accepted that man could survive very well in space for at least a couple of days.

The crew's role had undergone a major transformation from that of talking monkey to a vital, fully integrated part of the system. The word astronaut had become a totally understood and accepted term in the American vocabulary, and millions of kids aspired to become one. NACA's design had been bold and innovative, and McDonnell's engineering had been superb, building upon the best of aircraft technology (60% aircraft, 40% new, according to John Yardley). Perhaps it was true that the aeronautical branch of engineering attracted the best talent, as some claimed it had since the days of the Wright brothers. Certainly Gilruth, Faget, and the Space Task Group seemed gifted. About the only negative was that each Mercury capsule arrived at Cape Canaveral with a large number of defects that had to be corrected before flight. The flights themselves had been, overall, brilliantly conducted by the astronauts and their support team in Mission Control and tracking stations around the world.

Mercury was the last single-seat spacecraft, a sad thought for the fighter pilots of this world. Sure, people would fly solo in future spacecraft, as I did around the moon in 1969, but Mercury was the last single-seater. "Mercury" was also probably a misnomer because he was the *winged* messenger of the Gods, and the Mercury capsule had for good reason avoided wings. Perhaps Janus, the God of gates and doorways, would have been more appropriate. Janus is usually depicted with two faces, and certainly Project Mercury was a gateway, with its old face looking back toward the Wright brothers and its new one pointed directly at the moon.

When the Mercury program started taking shape, I was at Edwards, a student in the Test Pilot School. I would have given my eyeteeth to be part of it and later, as a fighter test pilot, I followed each flight with growing fascination. I wanted to fly higher and faster, to Mars even. To me that was more important than whether I landed with dignity on a runway or splatted ignominiously on my back in the ocean. There was one thing, however, that was disquieting: the Mercury capsule was so ugly! A small point, perhaps, but on the tarmac outside my window was a row of supersonic fighters. They were beautiful, exquisite aluminum sculptures that—even parked—looked as if they were going 1,000 mph. The Mercury shape, on the other hand, didn't have the grace of a thimble. It looked more like a trash basket. Furthermore, its blackened, corrugated surface gave it a rough, scaly look. Compared to sleek, flush-riveted aluminum, its bolted-on beryllium shingles seemed crude indeed. And as a final disgrace, it was topped with an escape rocket that looked like a lumpy parody of the Eiffel Tower.

Like an architect, an engineer wants his creation to look good, to be an aesthetic success, as well as being a visual expression of efficiency, a balance between weight and safety. The Mercury capsule looked like nothing before it, but did Faget's breakthrough in form have to seem so . . . so coarse? Perhaps it really was a jewel, a "sweet little bird" as Schirra put it, and heavens knows it was a tidy, tight package, but to those who designed and built it, was it pretty? Years later, I asked some of them. Faget was noncommittal: "Well, I guess so." Caldwell Johnson, who had sketched it over and over again, was more positive: "Yes, it looked pretty. I liked it because it struck me as being functional."

Scott Carpenter awaiting recovery

John Yardley gave the ultimate definition of the form-follows-function school: ''Pretty is what works.''

But it took Bob Gilruth for me to understand finally why Mercury *was* pretty. ''No, it wasn't pretty like a flower or a tree. But it had no bad traits. It was designed as a vehicle for a man to ride in, and circle the earth. With its blunt body, its retrorockets and parachutes, it was an elegant solution to the problem.''

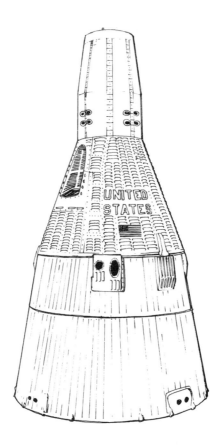

3

The Bridge

In the previous chapter, for the sake of continuity I omitted some key events external to Project Mercury itself but which impinged mightily upon it. During its latter phases Mercury was punctuated not only by Soviet successes but also by shifts in U.S. national policy as President Eisenhower was replaced by John F. Kennedy. Eisenhower had a space program thrust upon him by the Soviets and their Sputnik, and his administration never seemed comfortable with the new medium. Perhaps because of a "not-invented-here" feeling, there was no clear national space policy under Eisenhower. Although he supported NASA, and wanted civilian rather than military control over the manned program, Eisenhower finessed the question of where NASA should proceed after Mercury. When Glennan and Dryden asked for approval of a manned mission to the moon, shortly before he left office, he turned them down.

During the campaign of 1960, Republican Richard M. Nixon seemed to echo Eisenhower's reassurances that there was no emergency involving space, that the country should pull with an even strain. The Democratic candidate John F. Kennedy, on the other hand, seemed intrigued and excited about space and spoke frequently about it: " . . . we cannot run second in this vital race. To insure peace and freedom, we must be first. . . . This is the new age of exploration; space is our great New Frontier."

Kennedy's rhetoric was quickly put to the test. Within 3 months of his inauguration, Yuri Gagarin made the first manned space flight. A week after that, in the Bay of Pigs fiasco, Cuban defenders under Fidel Castro decisively defeated invaders backed by the CIA. April 1961 was the nadir of Kennedy's young administration. Stung by these reverses and looking for a badly needed fresh start, Kennedy wrote Vice President Lyndon B. Johnson, chairman of the National Space Council, a memo asking for an overall survey of the nation's space posture. He enquired specifically about the chances of beating the Soviets to the moon, by manned or unmanned vehicles, and he wondered about costs and the level of effort being expended: "Are we working 24 hours a day . . . if not, why not?"

Lyndon B. Johnson was a real space buff. As Senate majority leader, Johnson had chaired a special committee on space and astronautics that had a powerful influence on the final version of the National Aeronautics and Space Act of 1958. For years he had been urging the Eisenhower administration to be more bold in space.

While the Vice President, the new NASA Administrator, James E. Webb, Secretary of Defense Robert McNamara, and their respective staffs were pondering President Kennedy's questions, Al Shepard's successful Mercury flight on May 5 burst on the country like a breath of fresh spring air. Buoyed by the nation's exultant response to Shepard, and pleased to find that, in this case at least, NASA and the Defense Department agreed, Johnson recommended to the President that a manned lunar landing program be undertaken. He also recommended more satellites for communications and weather observation. Kennedy bought the package, and on May 25, in a speech to the Congress on "Urgent National Needs," the President outlined the program.

"First, I believe that this nation should commit itself to achieving the goal, before this decade is out, of landing a man on the moon and returning him safely to earth. No single space project . . . will be more exciting, or more impressive . . . or more important . . . and none will be so difficult or expensive. . . . It will not be one man going to the moon . . . it will be an entire nation." He asked Congress for an extra half billion dollars in the fiscal year 1962 and received it with remarkably little debate.

Thus the incredible commitment was made to journey a quarter of a million miles out into space, to land on a strange planet, to undertake an 8-day round-trip, when this nation's entire spaceflight data base consisted of one puny 15-minute foray, Al Shepard's healthy body, and photographs of a smiling Russian. Radiation, meteoroids, weightlessness be damned! We were going for the home run, all the way to the moon, to walk on its surface, and to do it all before the end of 1969! Kennedy had decided that it was in the national interest, the Congress agreed, and that was it. Kennedy had a lot of guts.

Another development that greatly affected the status of space exploration was the transformation of Bob Gilruth's Space Task Group at Langley into the Manned Spacecraft Center at Houston, Texas. This move took place in late 1961, after Enos' flight, but before Glenn's. The STG had grown to about 750 people, most of them wrestling with the day-to-day problems of orbiting a man, but as far back as 1959 Gilruth had been able to assign a few people to study advanced missions. The most popular of these were space stations visited by smaller craft, and trips to the vicinity of, or the surface of, the moon. So when President Kennedy said "a man on the moon," NASA had already done considerable homework on the problems involved.

As usual, the biggest problem was rocket power. A *huge* booster, dwarfing the Atlas, would be required for a lunar landing mission. In 1958 the Air Force had awarded a contract to the Rocketdyne Division of North American Aviation to begin developing the rocket engine for such a monster. This work, later transferred to NASA, resulted in a 1.5 million-pound thrust chamber, and engines that could be clustered to produce multiples of this number. How many for a moon landing? No one knew, but as studies progressed NASA discovered there were some interesting options.

The obvious method was to build one machine, a brute that would power its way from the earth's surface to the moon by direct ascent, slow down for landing, and then blast back to earth directly from the lunar surface. It would shed pieces as it went, but it was one machine. This was called Direct Ascent Mode. But clever engineers calculated that a machine that could separate itself into parts would be better. Launch two pieces independently, for example, join them in earth orbit and then proceed to the moon. This was called the Earth Orbit Rendezvous Mode, or EOR. Another option was to keep all the pieces together until in orbit around the moon, and then separate a landing craft that would return to the orbiting mother ship after exploring the surface. That was Lunar Orbit Rendezvous, or LOR.

Both EOR and LOR had their champions. Wernher von Braun's group at

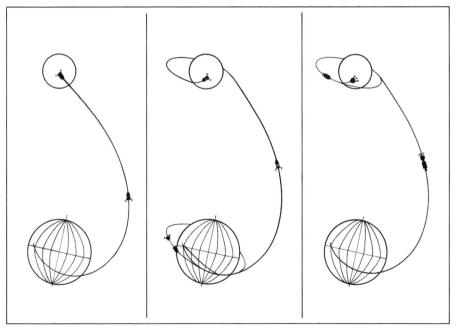

Direct Ascent   *Earth Orbit Rendezvous*   *Lunar Orbit Rendezvous*

Huntsville had been renamed the Marshall Space Flight Center and transferred from the Army to NASA. Marshall preferred EOR. Langley, spurred on by some innovative work by William H. Michael and others, and promoted by John Houbolt, was beginning to favor LOR. But in any event, it appeared wise not to have to depend on Direct Ascent and some huge future rocket, Nova, as it was being called.

The only problem was that both EOR and LOR assumed that the pieces could be assembled reliably in space, that is, they depended on what first was called orbital operations but soon became known simply as rendezvous. Rendezvous was absolutely vital, especially in the case of LOR, because a rendezvous failure around the moon would leave two astronauts stranded forever in space. Yet no one had ever conducted a space rendezvous, and it was no trivial task. Airplanes routinely found one another, most of the time at least, but in space things were a lot different.

You couldn't say "meet me over St. Louis" because an earth orbiting spacecraft has to keep moving at 5 miles per second. Furthermore, the branch of physics called orbital mechanics dictated some strange new rules for space pilots following curved paths. For instance, to overtake an object ahead of you in space it may first be necessary, against all piloting instincts, to fire thrusters pointed not toward the target, but away from it, so as to drop down into a lower and eventually faster orbit, and catch up that way. Seat-of-the-pants flying would not suffice for rendezvous; it would require radar, an inertial platform, a digital computer, and fuel—probably lots of fuel.

In planning the lunar landing mission, rendezvous became the most critical piece of the puzzle as far as the engineers were concerned. The medics had a different perspective: the trip time was 8 days, Gagarin had stayed up less than 2 hours, and no one knew what happened to the human body in weightlessness for a longer period.

While one small part of Gilruth's organization was preoccupied with these concerns, another part—initially just one man—was pondering what could be done to stretch the Mercury program to help provide some answers. Jim Chamberlain, the leader of the Canadian STG contingent, had moved temporarily to St. Louis to study the handcrafted capsule at its birthplace, to put his stethoscope on the fetal Mercury and prescribe a regimen for producing a more robust infant.

The heart of the problem was that Mercury had an exoskeleton. Everything was jammed inside ("stacked like a layer cake," as Chamberlain put it) so that equipment check-out and repair was a very tedious process involving the removal of healthy components to get at the sick one. Not only was equipment accessibility a problem, but the very nature of the exoskeleton design meant that growth in size or weight was very difficult. Mercury was already as densely packed as possible. For ease of manufacture and check-out, equipment modules should be externally mounted; in other words, a whole new design was required.

Furthermore, the complexities of rendezvous and the strain of long duration flight might require *two* crew members and there was no way to add an extra man to the Mercury capsule.

A growth version of Mercury was therefore going to have to be a brand new machine, designed from scratch with accessibility, ease of manufacture, operation, and maintenance in mind. The Mercury capsule had been brilliant in terms of making maximum use of very limited weight and volume, but it had gone as far as it could go. Of course the basic Mercury shape, heatshield, and parachute concepts could be retained, but Mercury Mark II, as it began being called, would have to be considerably larger to carry the extra fuel and accessories required for rendezvous and the extra expendables to sustain two crew members for a really long flight, perhaps as much as 2 weeks.

If Kennedy had said fly *around* the moon, instead of land on it, the basic Mercury's design could have been stretched enough to sustain one man on a 6-day circumlunar voyage—but a landing! A landing led inexorably to the untested concept of rendezvous, which in turn led to a two-man crew maneuvering in space with the assistance of on-board radar and computer. A landing, and subsequent exploration, also required the development of equipment and techniques for operating outside a spacecraft. And, of course, there was the question of the human body being able to sustain weightlessness for the 8 days required for a landing mission. The Mercury Mark II spacecraft was needed to answer these three questions of rendezvous, long duration, and extravehicular operation.

Jim Chamberlain and Bob Gilruth were able to convince NASA Headquarters that Mercury Mark II was a necessary gap filler between Mercury I and a lunar landing, and $75 million of the 1962 funds were earmarked for the project. Robert C. Seamans, Jr., NASA's Associate Administrator, had been recruited from RCA, where he had been working on an unmanned satellite interceptor for the Air Force. Because of this experience Seamans was very familiar with the concept of rendezvous, and he understood clearly the advantages of applying it to a lunar landing mission.

With the ink no sooner dry on Seamans' note of approval, Gilruth announced the news of Mark II, in Houston, to a wildly enthusiastic Chamber of Commerce audience. In January 1962, NASA released artists' sketches of the new Mark II spacecraft which, after an informal competition, it renamed Project Gemini. Congress had not been asked for its approval because NASA had the authority to switch the $75 million seed money from one research and development account to another, but anything designed to help meet President Kennedy's tight schedule was generally looked upon with favor.

Project Gemini got off to a jack rabbit start because it grew so naturally out of Mercury. Mercury's manufacturer. the McDonnell Aircraft Corporation, was the logical choice to build Gemini, and a contract was negotiated with the firm without taking competing bids. As the program developed, however, it became obvious that Gemini would have components totally unlike anything on board Mercury. Its booster would be a second generation ICBM, the Titan II, instead of the Atlas. But the Atlas was still required, this time to launch Gemini's rendezvous target, the maneuverable Agena rocket. The Gemini's nose would be fitted with a docking probe so that once Gemini and Agena found each other in space, the two vehicles could be joined together as a rigid unit and then the Agena's motor fired to propel the docked pair into a different orbit.

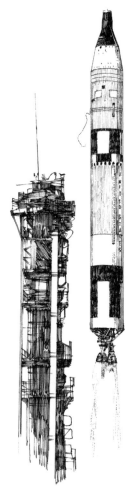

Titan II rocket

Radar and a digital computer were added to assist the rendezvous, and the ECS had to be modified to accommodate flights as long as 2 weeks. A novel power source, the fuel cell, was to replace the battery system, which was too heavy to provide electricity for long flights. An adapter section, a storehouse located aft of the heatshield, would hold extra supplies and equipment needed in orbit but which could be jettisoned prior to entry. The adapter also contained extra fuel so that Gemini could perform extensive rendezvous maneuvers.

A more mobile pressure suit would be required for extravehicular forays, not to mention a separate source of oxygen and cooling for an astronaut out on the end of a 50-foot umbilical cord. Two ejection seats would replace Mercury's escape rocket. Unlike Mercury, the Gemini spacecraft would generate some aerodynamic lift during entry, and could be steered to a precise landing. Finally, the landing itself would be on dry land, using a revolutionary new device, an inflatable wing, to both slow the descent and flare for landing like an airplane. Naturally where there are wings, there are wheels, so landing gear was also required. Taken all together, these changes from Mercury were enormous and had to be done very quickly, if a first flight were to take place in 1963, as Jim Chamberlain forecast.

The Titan rocket, as its name implies, is an object of great size and power. Almost 100 feet high, its two first stage rocket motors produce nearly half a million pounds of thrust, and its single second stage engine 100,000 pounds. Together the two stages had more than enough power to put an 8,000-pound Gemini spacecraft into orbit. The Titan, like the Atlas, had been developed by the Air Force as an ICBM, but its basic design was much improved over the

Atlas in terms of reliability, and Gemini was to be launched by an advanced model called Titan II.

The big change in the Titan II version was the selection of hypergolic fuels. A hypergolic fuel ignites spontaneously when mixed with the appropriate oxidizer. In this case, the fuel selected was a blend of plain hydrazine and unsymmetrical dimethyl hydrazine, and the oxidizer was nitrogen tetroxide. Hypergolics offer two advantages: simplicity and safety. The simplicity derives from the fact that no ignition circuitry or spark plugs are required. The safety is due to the fact that should fuel and oxidizer accidentally come into contact with each other, they will start burning right away, but in a less violent manner than the Atlas' kerosene and liquid oxygen, which explodes into a gigantic fireball. Thus, the smaller conflagration that would be produced by a ruptured Titan II allowed the crew to use ejection seats, instead of an escape tower, to vacate the premises. A final advantage of hydrazine and nitrogen tetroxide is that they are both stored at room temperature, so that no complicated chilling equipment is required, as it is in the case of liquid oxygen.

Although Gilruth and the others at first were pleased that Gemini had outgrown the temperamental Atlas, the new Titan II was not without its problems. Chief among them was something called the Pogo. The word Pogo was not an abbreviation, but was what engineers thought the problem resembled. The first stage of the Titan II vibrated longitudinally, so that someone riding on it would be bounced up and down as if on a pogo stick. The vibration was at a relatively high frequency, about 11 cycles per second, with an amplitude of plus or minus 5 Gs in the worst case.

This was below the threshold of visceral pain but it would prevent an astronaut from seeing his instruments and was deemed totally unsatisfactory by managers and crew alike. Titan II had some other problems as well, such as low thrust engines in the second stage. During 1962 and 1963 this supposedly mature, second generation ICBM produced very spotty test results, and only half of its first 20 flights could be considered fully successful.

Of all the changes from Mercury, probably the most radical was the return to "wings and wheels." The wing was unusual, to say the least. It was packed away like a parachute until the spacecraft had descended to 60,000 feet, at which point an elaborate unstowing and unfurling process began. By 20,000 feet, if all went well, the Gemini would become the world's heaviest hang glider, suspended under a dart-shaped, inflated fabric wing, or parasail. This concept was called the paraglider.

Like the Titan II, the paraglider ran into unexpected difficulties during 1962 and 1963, and more than once test vehicles were destroyed as they slammed into the ground. The reason for the paraglider was to allow the spacecraft to glide to a landing on a conventional runway, but from 20,000 feet it couldn't glide far, so the runway had to be close by. That, in turn, implied that retrofire had taken place precisely, and that the spacecraft had flown a predictable path during entry. Flight experience from Mercury, however, indicated that with a ballistic, zero lift capsule there would inevitably be dispersions which would prevent a pinpoint landing. In other words, without lift, if the spacecraft began to drift off course during entry there was no way to make a correction to bring it back on target.

If only Gemini could produce *some* lift, just a small amount, it could be steered during entry to that runway, so that when the parasail wing popped out it would be a simple matter to maneuver around into the wind and land like a conventional airplane, or at least a conventional glider.

Two things had prevented Mercury from generating lift: its shape, and a center of gravity located in the middle of its heatshield. Big brother Gemini had the same basic shape, designed primarily with entry heating in mind, but now Faget and his people did something they had deliberately avoided on Mercury: offset

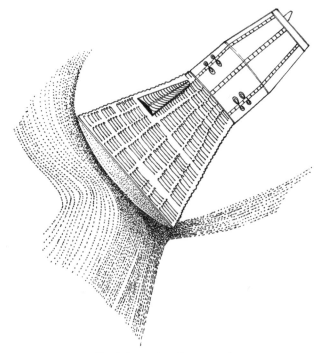

Gemini entering the atmosphere

the center of gravity. The spacecraft was built to have a symmetrical shape just like Mercury, but they loaded the heavy equipment into one side and the light equipment into the other, so that when air began flowing past the heatshield during entry, the spacecraft would tilt a little bit and enter the atmosphere cocked at an angle, with one edge of the heatshield slightly in front of the other. Thus the wind would strike it a glancing blow, and the Gemini would react by veering off.

In this simple way, a small amount of lift was generated without altering the external shape of the vehicle. The pilots could now steer toward a target by rolling the Gemini to whatever angle pointed the lift vector in the desired direction. This could have been done on Mercury but at that time Faget considered lift an unnecessary complication and something to be avoided. Besides, the Mercury astronaut was more a passenger than a pilot, and consequently the angular measuring equipment and extra maneuvering fuel were not provided. Another advantage of the offset center of gravity method of generating lift was that when the spacecraft was strapped to the Titan II during ascent, the slight imbalance of weight was negligible, and there was no tendency for the Gemini/Titan combination to veer off to one side.

Other changes to the Gemini were less radical than the parasail but more noticeable than the center of gravity shift, such as replacing the escape tower with ejection seats. Jim Chamberlain wanted to make this change but was dubious about what might happen if a Titan II exploded on the launch pad. In order to escape the expanding fireball, the seats would have to propel the crew about 300 feet up and 1,000 feet off to the side, a violent rocket ride in itself. The rocket catapult, called a rocat (NASA loves to create new names), had to be powerful, but a more demanding requirement was that it had to be aimed precisely through the center of mass or else man and seat would tumble ass over teakettle and foul the parachute when it tried to open. The parachute had to be fully open before the astronaut hit the ground, but it couldn't open too high lest an unfavorable wind blow him back into the fireball. The low altitude end of the ejection spectrum was the worst problem, but even at high altitudes things were hairy, as the seat tended to spin violently during a freefall through the thin upper atmosphere. To combat this, a small hybrid balloon-parachute called a ballute was deployed as a stabilizer. Rocats and ballutes! At this point the Gemini program seemed to be inventing words faster than solutions.

Since the main objective of Gemini was to prove the concept of rendezvous, it was natural that great care and attention went into the planning of, and the hardware for, the rendezvous missions.

The target vehicle, launched by an Atlas, would be the Agena, a workhorse used by the Air Force as an upper stage. Five feet in diameter and 25 feet long, the Agena was a flying gas tank with a rocket engine on one end. To the other end would be added a docking collar, into which the nose of the Gemini could be inserted. Once the two vehicles were locked together in orbit, commands could be sent from the Gemini to maneuver the Agena in any direction desired and to ignite its 15,000-pound thrust engine.

Like the Titan II, the Agena used hypergolic fuel and oxidizer, in this case unsymmetrical dimethyl hydrazine (UDMH) and inhibited red fuming nitric acid (IRFNA). One peculiarity of the Agena was that as far as we astronauts were concerned, the engine was on backwards. Instead of being pushed back in our ejection seats, as during launch and entry, the 15,000 pounds of thrust would jerk us out of our seats and plaster us against the instrument panel, unless we

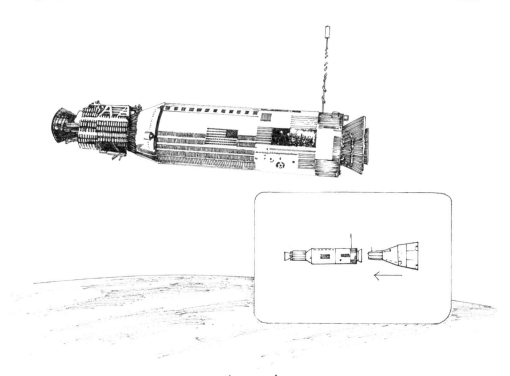

Agena rocket

were tightly strapped in. Despite this minor aberration, however, we really liked the idea of docking with the Agena. Unlike rendezvous, docking was a 100% manual operation, all eyeballs and seat-of-the-pants. It was quite similar to the mid-air refueling of a jet, and it put an end to the argument that spaceflight was totally automated, and that astronauts were along just for the ride.

If docking was straightforward, rendezvous was just the opposite. It was full of dark mysteries and our experience as fighter pilots intercepting each other did not seem to apply. Sir Isaac Newton, when formulating the laws of gravity and motion, had no idea how difficult he was making it for those of us who would fly his circles and ellipses. It was simple enough to explain, with a chalk board or, better yet, a powerful digital computer, but inflight one had to be extraordinarily careful not to make a false move, not to trust the eyes alone, not to fire the engines unless each maneuver had been checked and double-checked.

Consider, for example, a Gemini and Agena in the same orbit, with the Gemini a couple of miles behind and wishing to catch up. The pilot sees the Agena's twinkling light out the window, points the nose of the Gemini at it, and fires a thruster to move toward the Agena. For a short time all seems well, and the Agena grows in size. Then a strange thing happens: the Agena begins to sink and disappears under the Gemini's nose. Then minutes later it reappears from below, but now it is going faster than the Gemini and vanishes out front somewhere. What has happened? When the Gemini fired its thruster it increased its velocity but also its centrifugal force, causing its orbit to become larger. As it climbed toward its new apogee, it slowed down, so that it began to lose ground compared to the Agena. The Gemini pilot should have fired a thruster to move *away* from the Agena, causing him to drop down below it into a faster orbit, and begin to overtake it. Then, when the Agena reached a precisely calculated angle above him, he could thrust toward it and his resulting orbit would intercept that of the Agena. Sir Isaac demands that you play the game his way. The case I have described is, of course, a very simple one.

Where orbits of different shapes are considered, and orbits which are not in the same plane, the situation can become very complex indeed. Throw in some other variables, such as difficult lighting conditions or the requirement to bring the orbits together in a hurry, and great quantities of fuel can vanish in a futile chase. On the other hand, once two vehicles are within about 100 feet of each other in the same orbit, the effects of the curved path become negligible, and the fighter pilot's eyeballs can take over and complete the docking.

Essential to the success of a rendezvous is a precise knowledge of where the target is with respect to the overtaking spacecraft. Radar is used to determine this, in this case with the radar set on board the Gemini and a radar beacon called a transponder placed in the Agena target vehicle. The radar sends out an inquiring signal that is answered by the transponder. A computer then measures the time it took for the round-trip, at the speed of light, and from this calculates the range between the two vehicles, which is presented to the pilot and also recorded.

When this process is repeated, the two transit times are compared, and from them is deduced the rate of change of range, i.e., the closing rate (or opening rate, if the number is negative). Range and range rate are vital bits of information for the pilot, who must compare them constantly and keep them in the proper ratio so that he doesn't close too fast, and whiz by the target, or—just as bad—allow range rate to decrease too rapidly, and then have to waste extra fuel to speed up again. In addition to range and range rate, the radar measures the direction of the target relative to the spacecraft.

In order to measure direction, a fixed base for measuring angles in space is required. This ''stable table'' is generally referred to as an inertial measuring unit, and consists of three gyroscopes mounted at right angles to each other. From the angular movement of the spacecraft relative to these gyroscopes, pitch, roll, and yaw angles are derived, and from them can be calculated radar pointing angles. Three accelerometers are mounted in tandem with the gyroscopes, and they measure the spacecraft's reaction to the firing of thrusters as maneuvers are made.

With radar, an inertial measuring unit, and extra fuel for changing orbits, the Gemini pilot is almost ready for rendezvous. He needs one more thing, a computer containing a mathematical game plan describing the arcs in the sky that are best suited to bring Gemini and Agena together as expeditiously and economically as possible. Generally the shorter the intercept time, the more fuel it costs, so that the rendezvous sequence is a compromise between a short duration, brute power approach and a more leisurely series of maneuvers taking greater advantage of Sir Isaac Newton's laws.

On Gemini the "consumables" (the fuel tanks, oxygen for 14 days, and the electrical generating equipment) were all located in the rear adapter section, which was jettisoned prior to entry. The adapter also contained the maneuvering and attitude control system, whose fuel comprised 1,000 pounds out of a total spacecraft weight of 8,000 pounds. Thrust was provided by 16 small rocket motors that could be fired in pairs to control *attitude* (roll, pitch, and yaw angles) and to *translate* the spacecraft, that is to move it forward, backward, up, down, or to either side.

The pilot flew the Gemini in much the same way he would an airplane, with the stick, or attitude controller, in one hand and the throttle, or translational controller, in the other. For example, when approaching the Agena for docking, the commander's left hand manipulated the throttle, by pushing it up-down and left-right, to bring the Gemini onto the long axis of the cylindrical Agena, while his right hand made sure he was aligned with it in roll, pitch, and yaw. Then his left hand controlled the closing rate by fore-aft movement as he gently guided the docking probe into the Agena's collar. It was quite similar to flying a jet fighter's refueling probe into a tanker's drogue, except that in space there was no air turbulence from the Agena to rock the Gemini as it approached.

We copilots who were former fighter pilots had one added difficulty. There was only one attitude controller, mounted between the two seats. There were two identical throttles, one far left, one far right. Therefore, the man in the right seat had to train his hands to reverse roles, the left controlling the stick and the right the throttle. Even left-handers like me had some trouble with this at first.

The adapter section also contained an oxygen tank whose size depended on mission duration. Compared to maneuvering fuel, not much oxygen was required, only 100 pounds being provided for a short rendezvous mission, and 180 pounds for 2 weeks. Men breathe 2 pounds of oxygen per day, so two of them in 2 weeks only use $2 \times 2 \times 14 = 56$ pounds. But the spacecraft cabin leaks slightly, roughly 24 pounds being lost in 2 weeks. So that's 80 out of 180. The remaining 100 pounds was consumed in the production of electricity.

On Mercury the electrical power was produced by batteries, but they were too heavy to provide enough electricity for 2 weeks. To replace them a new device called the fuel cell was introduced.

Remember the old high school science class demonstration, in which an electrical current passed through water causes the H_2O to separate into a bell jar of hydrogen and another of oxygen? The fuel cell reverses this process by sustaining a chemical reaction that combines hydrogen and oxygen to produce electricity and water. Not only is this system lighter than equivalent batteries, but the water produced as a by-product can be used as part of the crew's drinking supply. A man drinks about 6 pounds of water a day, or 168 pounds for two men for 2 weeks. A hundred pounds of oxygen combining with 20 pounds of hydrogen could fulfill nearly three fourths of this requirement.

Unfortunately, in the mid-sixties the fuel cell was still an experimental device and the Gemini cells not only had a short operating life, but the water produced by them turned out to be the color of strong coffee, contaminated by organic particles we called "furries." It was undrinkable. Even so, weight was saved by using the cells, and as crew members we became quite attached to them. Unlike batteries, they had character. There were six of them, arranged in two units of three, and no two of them performed exactly the same way. Like sled dogs, some were stronger and pulled more than their share of the electrical load; others were malingerers and had to be coddled and rested periodically.

Shortly before each launch, the crew was handed a piece of paper listing the latest state of health of each cell, in the form of a graph showing how many volts it was then capable of producing at light and heavy loads. In flight, actual voltmeter readings were compared with predicted output, to warn of impending failure. Also, every 4 hours or so the fuel cells had to be cleansed, or purged as it was called, by force-feeding them extra hydrogen for 13 seconds and oxygen for 2 minutes. For me, 13 seconds was no problem, because I kept my finger on the switch, but I usually did a couple of other things during the 2-minute oxygen purge and forgot to turn it off. Fortunately, unlike overfeeding sled dogs, no harm was done.

Provisioned with a 14-day supply of oxygen, hydrogen, water, food, and lithium hydroxide (to absorb carbon dioxide), the Gemini spacecraft could sustain a flight duration almost twice that of a lunar round-trip. By adding extra propellant (hydrazine and nitrogen tetroxide), plus radar, computer, and inertial measuring unit, it could be used to demonstrate various rendezvous schemes that might be required around the moon. Gemini could not be used to duplicate walking on the moon in $\frac{1}{6}$ G, but it could begin to investigate some of the problems of extravehicular operation, and develop some of the equipment needed. Out of this requirement grew the Gemini "space walks."

First on the equipment list was the spacesuit, or pressure suit as we astronauts called it. In the U.S. the pressure suit traces its roots back to Wiley Post, the one-eyed pioneer best known for perishing together with Will Rogers in a 1935 airplane crash at Point Barrow, Alaska.

In 1933 Post had been the first to fly solo around the world (8 days in a Lockheed Vega), and the flight gave him a yen to fly higher and avoid much of the weather. The B. F. Goodrich Company built him a pressurized rubber suit with a bolt-on aluminum helmet that had the faceplate offset slightly to center itself on Wiley's one good eye. It looked like a deep-sea diver's outfit. Post was killed shortly afterwards, and pressure suit technology languished until after

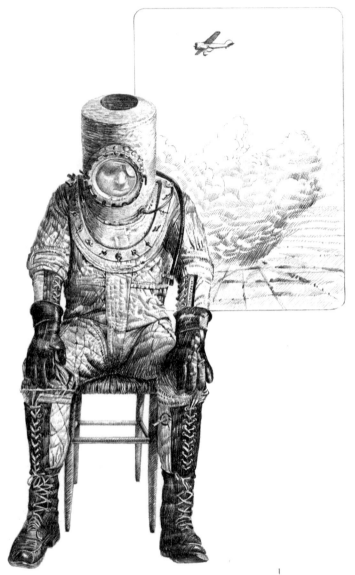

Early pressure suit . . . Wiley Post

World War II, when aircraft routinely began to exceed 50,000 feet. Above this altitude there were cases of severe bends, and pilots were getting perilously close to the 63,000-foot height at which body fluids begin to fizz and bubble.

A modern pressure suit looks deceptively like a slightly modified flight coverall and helmet combination, but beneath that fabric is some highly imaginative engineering. The fundamental challenge is to design something nearly skintight that will not balloon badly when inflated, and is sturdy enough to endure harsh treatment, temperature extremes, and meteoroid impact, yet at the same time retain its comfort and its flexibility.

Gemini pressure suit . . . Ed White

Flexibility versus structural integrity; comfort versus restraint— the tradeoffs
are very difficult. For example, if the helmet sits comfortably on your head
before the suit is inflated, it may rise up as pressure inside increases, until you
can only see straight up. To prevent this stretching, the designer can add
restraining straps, but when the pressure is released they may dig into the tops
of your shoulders and cut off your circulation. Or consider a more complicated
problem: the shoulder joint. The arm rotates at the shoulder, swings up and
down and left and right. How to design a pressure-tight, comfortable,
nondeforming joint that will still allow the arms to perform all these motions

through their normal range and without having to exert undue effort? Like an automobile inner tube, an uninflated suit is limp, flaccid, but as the pump is put to it, and the pressure inside it builds, it becomes rigid, frozen in one shape. To cause it to deviate from that shape, as the human body inside it must in order to perform useful work, joints must be provided at wrists, elbows, shoulders, hips, knees, and ankles. (Suit designers had given up on the waist as too difficult a problem and allowed it to remain fixed. That's why you never see a space-walking astronaut bend over, bow, or curtsy.) These joints also have to be free and easy in their motion, or as the engineers put it, they must be low-torque.

Various schemes have been tried to assure low-torque joints, using bellows, pulleys, and cables. The Gemini suit relied on a fishnet-like, one-way stretch, loosely woven fabric called Link Net. Link Net's great virtue was unpressurized comfort obtained at the sacrifice of pressurized mobility. You had to work hard to bend a pressurized Gemini suit. The Apollo moon suit was more mobile but not as comfortable. The early ones, especially, dug into shoulders or crotch, or both.

In addition to sustaining a life-saving pressure of 3.7 psi of oxygen on the body of a spacewalker, the Gemini suit also had to protect against temperature extremes and meteoroid impacts. Temperatures vary wildly in space, from plus 250° at noontime to minus 250° at midnight, and the interior of the suit had to be impervious to the bake-freeze cycle. Of the two temperature adjustments, keeping cool is the more difficult problem, and a network of tubes was sewn into the suit, to deliver cool oxygen to arms, legs, and torso. Ducts also carried oxygen to the helmet visor, to prevent fogging. The circulating oxygen carried away exhaled carbon dioxide, which was absorbed by lithium hydroxide contained in canisters in the Gemini's environmental control system.

A layer of felt was added to the suit to serve the dual purpose of keeping warm at night and protecting from meteoroid hits. In daylight, the sun's energy was reflected by aluminized mylar and nylon layers. When fully dressed and ready to go outdoors, the astronaut was protected by the following (from the inside out): (1) long cotton underwear, (2) a nylon comfort liner, (3) a pressure bladder of neoprene coated nylon, (4) a Link Net restraint layer, (5) one layer of felt, (6) seven layers of aluminized mylar superinsulation, and (7) a cover layer of high temperature nylon.

Helmet and gloves, which were removed when not needed, were attached to the suit by ingenious rings that locked together tightly but still rotated freely. The helmet had a visor that could be opened, plus an over-visor that served as a sun shield. Earphones piped in intercom and radio circuits. To be heard, the spacewalker had only to talk. A device called a VOX, or voice operated relay, sensed the noise coming into the microphone and triggered the appropriate circuits, either intercom only or intercom plus radio, depending on the configuration of the communications switches.

The astronaut donned and doffed the suit through the rear, with a heavy-duty zipper running from the back of the neck down to the crotch. Since the zipper, of necessity, penetrated the pressure bladder, some ingenious engineering was required on what is normally a trivial clothing accessory. Interlocking rubber lips on either side of the zipper formed a pressure-tight seal. Still, the notion of this zipper always bothered me and some of my fellow astronauts. To this day, astronaut John Young has a cartoon on his office wall showing a technician summoned to investigate a launch delay, shouting down from the top of the gantry: "He sez his zipper is broke!"

The gloves, an accessory in our terrestrial wardrobe, are an absolute necessity in space. In most cases they are the weak link in the extravehicular chain. To this day they obstinately refuse to bend (literally) to the most modern technology. When inflated they tend to become flat and bloated with fingers extended. To form a fist or to grip a tool requires effort, not much effort, but the torque is sufficient to cause the hands to tire before the rest of the body. The Gemini models had plenty of padding to enable the astronaut to hang onto hot objects, and delicate little battery-powered lights were sewn into the fingertips.

Other necessary accessories were boots (for once something simple that didn't require much design effort), three chest connectors for inlet and outlet oxygen hoses plus communications umbilical cord, and a chest-mounted device called a helmet tie-down strap, whose function it was to vary the distance from helmet to crotch, acting as a crude method of partially bending the waist.

Inside the suit were other odds and ends, such as biomedical sensors taped to the chest with wires running to electronic amplifiers placed in pockets in the underwear. Finally there was a "motorman's friend," a triangular urine bag with a condom-like device into which the astronaut inserted his penis before donning the suit. Once you were locked inside the suit, none of this gear could be adjusted, nor could an eye be rubbed, a nose blown, or an itch scratched. Some of the most fundamental amenities that we take for granted on earth are difficult or impossible for an extravehicular astronaut.

The Gemini suit was a natural extension of the Mercury suit, but there was one fundamental difference in usage and hence importance. On Mercury the suit was a back-up device in case the cabin pressure fell to a dangerous level. During Gemini extravehicular operations, the suit became the primary pressure vessel *with no back-up*. It was not practical to build any redundancy into it. If the visor popped open, if a glove flew off, if the rubber bladder ripped, you had a dead astronaut floating around on the end of an umbilical. There was no way for the surviving astronaut to stuff the body back into the cramped cockpit; it would have to be cut free to drift in space. Depending on the altitude, the dead man might orbit the earth for years.

Of all the "failure modes," as NASA calls them, to me the most worrisome was a pressure bladder leak. In order to fit each astronaut precisely, the bladder was a cut-and-paste job. Seamstresses at the David Clark Company were highly skilled and motivated, but still. . . . Like Caldwell Johnson's lament about those who sewed parachutes, suit construction seemed a haphazard process to tidy engineering minds, and some officials were not enthralled about the risks involved. Putting all your trust in glued joints in thin rubber fabric just didn't seem prudent. As John Young put it, NASA was not overjoyed about ''putting guys in vacuums with nothing between them but that little old lady from Worcester, Massachusetts, and her glue pot and that suit.''

Unlike the Mercury, which had a bolt-on hatch, the Gemini was provided with two hatches, either of which could be opened and closed repeatedly from within. Also, no two Geminis were exactly alike. Four of the manned flights planned no extravehicular activities. The other six carried a variety of gear. On Gemini 10, for example, I went out on the end of 50-foot umbilical. This is too great a distance to use the spacecraft ECS to provide low pressure oxygen or cooling. A chest pack was required that took high pressure oxygen from the umbilical cord and reduced it to 3.7 psi and fed it into the suit. The chest pack also contained a fan for circulating the oxygen and a cooling device, a heat exchanger that operated by boiling water.

My first encounter with a real Gemini was in late 1964, at the McDonnell plant in St. Louis. Partially assembled, it was undergoing tests in the clean (white) room. Today clean rooms are quite commonplace, thanks to the electronics industry. A speck of dandruff in a computer circuit is like a boulder in the middle of a highway. But in 1964 the clean room was an oddity, and the idea of using huge fans to suck impurities out of the air was new to me. I felt absurd putting on a white nylon gown, plus hat, gloves, and little white booties over my shoes.

I had never seen a Mercury, except on television, so this was my first peek at *real space machinery*. There were several Gemini spacecraft in the room and they were hard to make out at first, because they were sitting on their blunt ends with their cylindrical snouts pointed up, and they were surrounded by scaffolding, workbenches, test equipment, and other miscellaneous gear. They were like asparagus spears peeking up through a dense tangle of weeds. They were most noticeable not by shape but by their dull black color. They were not graceful, but they were business-like. They were serious machines, dormant now, but with a hint of power and speed in them. They looked strong.

But God they were small and uncomfortable! Was I sure I knew what I was letting myself in for? The back of the ejection seat was canted forward slightly, at an angle that seemed calculated to dislocate the stoutest sacroiliac. With legs up in the air and gravity plastering the torso back against the headrest, it was hard to imagine spending much time in this machine, certainly not the *2 weeks* that was being talked about. With the hatch closed, it was worse, much worse.

In a fighter cockpit you may be cramped, but the great expanse of Plexiglas overhead gives you a clear view, and a free, open feeling. The Gemini windows were tiny, and when the massive hatch slammed shut I felt locked in, a sensation

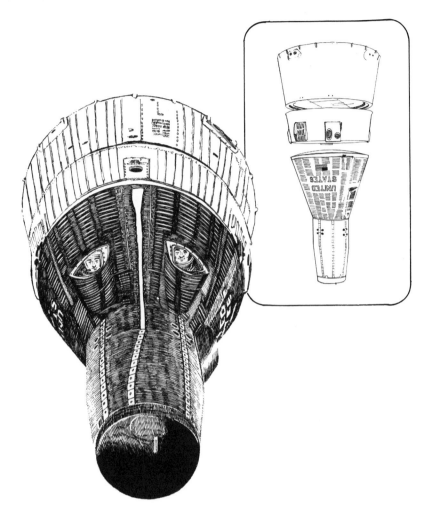

Gemini

I had never experienced in the smallest jet cockpit I knew, that of the Lockheed F-80. And all this in shirtsleeves. How would I feel when also crammed into a heavy, hot, bulky pressure suit? I knew the Mercury had been tiny, but this was just as bad, maybe worse. The diameter of the Mercury was 6 feet. Gemini's was 7½ feet, but cabin volume, instead of being twice as great, was only 50% larger.

They would need a shoehorn to get two of us, pressure suited, into this little jewel. At the time of this visit to Mr. Mac's clean room I had in my garage at home a garbage can painted flat black with two simulated windows on the side of it. It had been a going-away present from my Edwards test pilot buddies, and it kept popping into my mind's eye as I looked at the real McCoy. They weren't all that dissimilar.

I had been hired as a NASA astronaut in October 1963, between the Mercury and Gemini programs. I was one of fourteen, in the third group selected. The original seven were, without doubt, the Mercury astronauts. The group of nine that followed was sometimes referred to by the press as the Gemini astronauts, and we fourteen were dubbed the Apollo astronauts. But it wasn't that simple.

Three of the Mercury seven flew on Gemini, as did eight of the second group (four of them twice) and five from my group. In 1964 there was only one manned spaceflight, the first of the new Russian three-man craft called the Voskhod (this word does not translate easily into English. It means something close to "rising"). There were nearly 2 years between Gordo Cooper's Mercury finale and the debut of Gemini twins Gus Grissom and John Young.

I needed this delay, although I didn't realize it at the time, because I had so much to learn. I had begun, while still at Edwards, by attending a postgraduate course at the Air Force Test Pilot School, which at that time preferred the grander title of Aerospace Research Pilot School. There we studied the fundamentals of spaceflight, such as the mathematics of orbital mechanics, the backbone of rendezvous theory. We also studied physiology, in a deft survey course of the human body. If we understood how it worked on earth we might have a better appreciation of how weightlessness could affect its functioning.

So when I arrived at the Manned Spacecraft Center in Houston early in 1964, I had some fundamental training in astronautics but no specific knowledge about Gemini other than what I had gleaned from technical magazines. The clean room was a good place to begin.

NASA had originally hoped to launch Grissom and Young by the time I "came on board," but the program had fallen behind considerably. Most of the blame was placed on the Titan II and its problems of combustion instability and Pogo oscillations. Moreover, far from being the modified Mercury that Jim Chamberlain had initially hoped for, Gemini turned out to be a spacecraft re-

Chuck Mathews

quiring a lot of new technology that was slow to develop. By this time Chamberlain had been replaced as program manager by Chuck Mathews, who was deemed to be a steadier influence and a better communicator than Chamberlain, a brilliant designer who, however, tended to favor oversophisticated designs.

Chuck Mathews found that development of components such as the ejection seats were lagging badly and that one—the paraglider—seemed hopelessly snarled in a financial, technical, and managerial morass. Fortunately the paraglider was not absolutely necessary, being more technological frosting than cake, but it was nonetheless a nice touch to be able to glide to a runway, even though by this time Apollo was committed to a water landing. After a year of unsatisfactory progress, Mathews killed the paraglider in mid-1964.

My first assignment in Houston was to become the astronaut office specialist on pressure suits and extravehicular operations. That was a hopeful sign because I wanted to walk in space and on the moon, and I thought if I could help develop the equipment I might have an advantage over some of the friendly competition, my fellow fourteen. On the other hand, they had all been given technical specialty areas, too, and perhaps some of theirs (boosters, navigation) were more important than mine.

I spent 1964 learning the basics and some frills, too, such as how to cook iguana lizard in the jungle and how to survive with a minimum amount of water in the desert. I traveled around the country to the various space places and little by little I got comfortable with the idea that I might go up into space—and fairly soon at that. The only Gemini flight in 1964 was the first, Gemini-Titan I. Unmanned, GT-I was more a test of the Titan II than of the Gemini spacecraft, and it was very successful. Unlike the Atlas, which departed in a raucous cloud of yellow fire and white smoke, the Titan II was more demure, its hypergolic fuels emitting an almost transparent plume that NASA described as lambent—softly bright or radiant.

The pace of flights really picked up in 1965. There were seven altogether, the first unmanned and the next six each carrying two men. No apes made it into space on the Gemini program, a relief to us astronauts who detested Mercury's simian comparisons.

The crew of Gemini 3, Gus Grissom and John Young, were a matched pair. They were both good engineers who understood their machines and liked fooling with them. They were uncomfortable with the invasion of privacy the space program had brought into their lives, and tried as hard as they could to deflect questions from themselves to their beloved machines. They were generally taciturn but both had strong opinions that could flash unexpectedly and, in Gus's case at least, angrily. Neither was interested in small talk, and they would endure uncomfortable silences rather than fill the void with what they considered ancillary trivia. John is witty, but rarely allows himself to be perceived as such, especially when outsiders are around. He prefers a cloak woven partially of engineering mumbo jumbo and partially of aw shucks, t'ain't nothin', southern boy platitudes. Gus used no cloak. Both had a strong work ethic and kept long hours.

Acutely aware of the fact that Liberty Bell 7 rested on the bottom of the Atlantic Ocean, Gus named Gemini 3 Molly Brown, after the heroine of a Broadway play, *The Unsinkable Molly Brown*. NASA was not pleased with this frivolous choice but since Gus's back-up name was Titanic, they relented. It was a Pyrrhic victory, however, for the resistance of NASA's hierarchy to names for individual spacecraft had been building since Gordon Cooper's flight (what if we lost Faith?). Consequently, despite the fact that the crew of Gemini 4 wanted the red, white, and blue name of American Eagle, the custom of naming spacecraft was discontinued for the Gemini series and for the first Apollo flights. Of necessity it returned during the flight of Apollo 9, when the vehicle separated into two parts, the Lunar Module and the Command Module. During their subsequent rendezvous, Mission Control could not refer to both of them as Apollo 9, so the crew was once again permitted to bestow names. Their choices, Spider and Gumdrop, were wonderfully descriptive of the appearance of their craft, but again lacked the dignity sought by program officials.

Molly Brown's flight was for three orbits in 5 hours, long enough to give new equipment a fair shakedown but short enough to avoid any unnecessary risks. Gus and John had wanted an open-ended flight plan, wherein they could decide to stay up longer if everything looked all right, but Chuck Mathews took a more conservative approach, and his view prevailed. Like all program managers, Chuck had a long list of things that worried him and he felt that 5 hours was plenty for a first manned flight.

The most important difference between Molly Brown and the six Mercury flights that preceded it was the ability of the Gemini spacecraft to change orbits by firing its own thrusters. This was a fundamental requirement in the rendezvous process, and Gemini 3 demonstrated its ability to move about in space in three carefully calculated "burns."

The first maneuver was 75 seconds long and changed the orbit from elliptical to circular. The second twisted the plane of the orbit slightly, and the third, on the final orbit, lowered the perigee sufficiently so that the spacecraft would have entered the atmosphere even if all four of its solid retrorockets had failed to fire. During entry the crew guided their craft toward the touchdown point as best they could, but the offset center of gravity did not produce nearly as much lift as the wind tunnel had predicted, and they landed about 50 miles short. During the parachute deployment both men were thrown forward into the instrument panel and Gus's faceplate shattered. All subsequent crews were careful to put a shielding forearm up in front of their faces before touching the parachute switch.

Despite the fact that Molly Brown had proven to be remarkably trouble free, and that Gus and John had performed flawlessly, post flight attention was drawn to, of all things, a corned beef sandwich. This particular corned beef sandwich had been purchased by Gus's back-up, the chief astronaut prankster Wally Schirra, who gave it to John Young. When it was time for the crew to eat their scientifically designed, painstakingly packaged space food, John casually reached into his pressure suit pocket and offered the sandwich to Gus, who good naturedly took a couple of bites. News of this outrage leaked to the press and then to the Congress. At the next appropriations subcommittee hearing, NASA Administrator Jim Webb and 44 of his staff were gathered to testify. Billions of dollars hung in the balance.

Congressman Shipley: "My thought is that . . . to have one of the astronauts slip a sandwich aboard this vehicle, frankly, is just a little bit disgusting.''

Dr. Mueller (Associate Administrator for Manned Space Flight): "We have taken steps . . . to prevent recurrence of corned beef sandwiches in future flights."

Dr. Gilruth (trying to restore a little sanity to the proceedings): "These things do help to break up the strain."

Mr. Webb (having no part of *that*): "I do not agree that you can tolerate this kind of deviation. . . ."

As far as I know, no corned beef sandwich has ever again flown in space. Sliced turkey on whole wheat, maybe. . . .

Always mindful of the fact that the objectives of Project Gemini were long duration, rendezvous, and extravehicular operations (EVA), Chuck Mathews wanted to move as swiftly as he could, and Gemini 4 looked like a chance to work on all three objectives.

The first two hurdles to long duration were the fuel cells and the conservatism of the medics. Chuck wanted a 7-day flight but the fuel cells were still flunking their tests so, like its predecessors, Gemini 4 had to fly with battery power only, which meant a maximum of 4 days.

Even that was too long for the medics because, unlike in the Mercury, when the Gemini was floating in the water the crew was seated in an upright position. After a long duration flight, orthostatic hypotension caused pooling of blood in the legs, and this might be sufficient to cause the astronauts to lose consciousness when they returned to earth. If they passed out and fell down on the carrier deck, no problem, the blood would return to their heads and they would regain consciousness. But inside the Gemini, their torsos would remain strapped upright, blood would drain out of their heads, and the result, said the medics, might be two fatalities.

Mathews felt they were exaggerating: Gordon Cooper had merely been light-headed, not close to passing out, and the effects of inflight exercise should counteract, to some extent, cardiovascular weakening during weightlessness. Four days it was.

The second objective seemed simple: upon reaching orbit, the Gemini would separate from the second stage of the Titan II and then fly back to it—a miniature rendezvous in itself.

The third objective was EVA, or as the press dubbed it, a "space walk." A special chest pack and a hand-held maneuvering unit had been hurriedly assembled for this purpose.

The crew of Gemini 4 was Jim McDivitt and Ed White, both Air Force officers who had been classmates at the Test Pilot School. They were as extroverted as Grissom and Young had been introverted. Jim was extraordinarily bright, and had graduated number one in his class both in college and at the Test Pilot School. Ed was an all-American type, a wonderful athlete with a huge smile, and shared with John Glenn an unusual—for astronauts at least—willingness to speak without embarrassment about himself and the space program in terms of national goals and a personal sense of patriotism. Like Gus and John, McDivitt and White were also a matched pair—but of an entirely different sort.

Gemini 4's beginning was not an auspicious one. The Titan II second stage proved to be an elusive target for rendezvous. When McDivitt zigged toward it, it zagged away. He tried to approach it from several directions but was not able to get closer than 100 yards or so. After he had used up half his maneuvering fuel, McDivitt conceded defeat to the twisted laws of orbital mechanics and gave up the futile chase.

Next came the EVA, and it was a spectacular success, tempered only by— the Russians again! Not 3 months before, Cosmonaut Aleksei Leonov had dangled from the end of a cord attached to Voskhod II, the first human to venture outside the womb of an orbiting spacecraft. Nonetheless, Ed White approached his novel task with his customary enthusiasm and after a slow start with a balky hatch, he popped out in front of the spacecraft, where McDivitt took some magnificent pictures of America's first EVA.

The hand-held maneuvering unit, a primitive device that propelled Ed by squirting out jets of compressed gas, soon ran out of fuel, and much to his disappointment, Mission Control ordered him back inside. "It's the saddest moment of my life," he radioed. Again there was an anxious moment as the hatch resisted Ed's first attempt to shut it, and his athlete's slow pulse zoomed up to 180 beats per minute. After some hauling and tugging (McDivitt pulling on White's legs while White pulled the hatch down), the hatch closed and the spacecraft was repressurized. Ed was tired, had sweat in his eyes, and his visor was fogged over, but both men were exhilarated at the idea that at nearly 18,000 mph, Ed had been able to simply open the door and step outside.

The rest of the flight was anticlimactic, although the 4 days passed quickly because NASA had decided to load a variety of experiments on board the long duration Gemini flights. Dosimeters, spectrometers, cameras, magnetometers— these devices kept the crew busy photographing and measuring. Exercise, food preparation, and routine housekeeping chores were also time-consuming. Gear was stowed in deep, narrow compartments, and the item desired always seemed to be on the bottom of the stack. That meant a floating armada of unneeded gadgets circling the tiny cabin and bumping into each other. All these had to be scooped up and crammed back into their cabinets. If a compartment door was left open, out they all floated again.

Gemini 4's computer had not been programmed to help with the rendezvous, but it was supposed to provide entry steering commands, so that McDivitt could fly a precise trajectory and land next to the pick-up aircraft carrier. Unfortunately, the computer broke toward the end of the flight and McDivitt's seat-of-the-pants entry missed the boat by 50 miles.

Still, Gemini 4 was rightfully judged a great success and newspapers around the world followed its progress carefully. After the flight President Johnson himself came to Houston to congratulate Jim and Ed, whom he then sent to the Paris Air Show where they met with Yuri Gagarin.

With a 4-day flight under his belt, Chuck Mathews was ready to double up. So was the Gemini 5 crew, Gordon Cooper and Pete Conrad. "Eight days or bust" was their motto, and a covered wagon the motif of their crew patch. Cooper and Conrad were very different: Gordo was an uncomplicated, laconic Oklahoman while Pete was a witty, ebullient storyteller from the Philadelphia main line. But they were united in their desire to go the full 8 days. They knew, as did Mathews, that duration depended on those new-fangled contraptions called fuel cells. The fuel cells really bothered Chuck Mathews. Twenty years later he still recalls "One of the nightmares I had—an insurmountable fuel cell problem."

Sure enough, after two orbits of the Gemini 5 flight, Pete Conrad noticed that the pressure in the tank supplying oxygen to the fuel cells was steadily dropping. It should have been up at 850 psi, but gradually fell off to 70 psi. The crew powered down the fuel cells, an act that forced them to abandon a rendezvous exercise with a special pod they had carried into orbit and already deployed. The last time they saw the departing pod on their radar it looked fine, but they didn't have the power to chase it. However, the fuel cells were merely dormant, not destroyed, and as the flight progressed the oxygen pressure gradually recovered.

Because of this an 8-day flight was possible, albeit under somewhat reduced activity levels—toward the end even Conrad got tired of talking to the uncommunicative Cooper and wished he had brought a book along. One final indignity, landing 90 miles from the boat, was traced to the fact that someone on the ground had loaded the earth's rotation rate incorrectly into their computer.

It took Cooper and Conrad 2 days to bring their cardiovascular systems back to preflight levels, but these were just clinical observations, not noticeable to them, nor indicative of any impediments to longer flights.

Thus in five short months, in early 1965, the flights of Gemini 3, 4, and 5 provided Chuck Mathews with a sampling of EVA and a large dose of data on the physiological effects of weightlessness. Only the two rendezvous experiments had been disappointing.

The next flight, Gemini 6, was to concentrate on rendezvous alone. "We couldn't afford to play with experiments," said crew commander Wally Schirra, "rendezvous was significant enough." His copilot was Tom Stafford, a brainy member of the second group who had been an instructor of mine at the Test Pilot School. They were a good choice for the first rendezvous mission, Stafford's penchant for things electrical and mathematical acting as a strong right arm for the cool commander Schirra, pulling as usual with an even strain. They were the first crew to receive extensive rendezvous training in the McDonnell simulator at St. Louis, and by launch day they had completed over 50 practice runs.

Gemini 6 was the first flight to have an Agena target vehicle. The launch of the Gemini and Agena had to be carefully coordinated, so Schirra and Stafford were strapped into their cockpit as the Atlas belched flame and took off with their Agena perched atop it. The Atlas did its job, but when the Agena was released and ignited its own engine, it got a hard start—similar to an automobile's backfire—starting a chain of events that ended with the Agena being blown into five pieces. Disgusted, Schirra and Stafford climbed out of Gemini 6 and joined program officials in puzzling out what to do next. Rendezvous was proving to be more elusive than expected.

It soon became apparent that it was going to take months to analyze and cure the Agena's ills, but that need not delay the entire flight sequence. Gemini 7, next in line, was to be *the* long duration flight of the series and did not require an Agena. The crew commander was Frank Borman, whom I knew well, as we had been classmates at the Test Pilot School. I know no one more tough or tenacious, no one better suited to endure 14 days jammed with another person into a space the size of a phone booth.

His copilot, Jim Lovell, had been given the horrendous nickname "Shaky" by his buddy Pete Conrad, but he was anything but shaky—Lovell was an affable, smooth, steadying influence on the outspoken and irascible Borman. Frank's back-up was Ed White, and I was Jim Lovell's. I was really pleased to drop my pressure suit work and join a back-up crew, because that was the first step in the process of getting a flight of my own. The long duration flight was not the one I would have chosen, because we were more guinea pigs than pilots, but as the first of my group of fourteen to be picked, I was overjoyed to be on *any* crew. I had mixed feelings on December 4, 1965, as Gemini 7 headed east from Cape Canaveral on its long journey without me; sad that I wasn't on board, but tickled pink to be a step closer to my own flight.

No sooner had Gemini 7 departed than Launch Pad 19 became a beehive of activity as the ground crew readied Gemini 6 to follow it. Rather than wait for lengthy Agena repairs, Chuck Mathews had decided to go along with a McDonnell suggestion that Gemini 7 be used as a target in lieu of an Agena. The two Geminis could not dock with each other, but rendezvous was the main goal, and docking could be demonstrated later. With only one Gemini-Titan pad at the Cape, launch damage to it had to be repaired, and the new booster and spacecraft mated, erected, and checked out. It took 8 days and lots of overtime before Schirra and Stafford were ready to try again.

Their countdown was trouble free, and the two Titan main engines ignited on schedule. After 1 second the clock started, a clue to the crew that they were on their way. At the same instant, however, there was abrupt silence. The engines had stopped! Was Schirra airborne or not? If the Titan had risen from the pad even an inch or two, the massive booster was even now sinking back to earth, where it would crumble, fall on its side, and explode—not necessarily in that order. If it had not moved at all, it was still bolted firmly in place. There was a lifetime of difference between the two situations. If the Titan had moved, Schirra must make an almost instantaneous decision to eject, and yank the seat ejection ring between his legs. If there had been no motion, he should do nothing.

In training it had been emphasized that the clock could not start until a plug at the base of the gantry had been pulled loose by the departing Titan. Tick! Tick! There was the clock in front of Wally's nose. What did he do? Nothing. What did Stafford do? Nothing. What should they have done? By a strict interpretation of the rules, Schirra should have pulled the ring and ejected both of them. However, years of flying had taught Wally that sometimes you had better believe the seat of your pants, or more accurately, all the kinesthetic impulses feeding into the brain from eyes, ears, tendons, joints, and muscles.

This time Wally believed his body, and he was absolutely right. The tail plug had dropped loose prematurely and sent a false signal to the cockpit. The malfunction detection system had reacted by shutting down the engines. Cool as a cucumber, Wally sounded on the radio as if this kind of thing happened to him every day. Once again, however, the frustrated pair had to climb out of their Gemini while it was still anchored to Launch Pad 19.

As it turned out, Wally's decision was not only correct, it was pure luck he had been confronted with it, and that the plug had picked that particular booster to trick, because a routine examination later showed something mighty suspicious in the thrust build-up in one Titan engine. Sure enough, disassembly produced a plastic dust cover that had been carelessly left *inside* the engine. In all likelihood this engine could not have produced enough thrust to boost Gemini 6 into orbit.

Within 3 days everyone was ready to give it a third try. As Borman and Lovell passed overhead, off Wally and Tom went into a carefully calculated tail chase. After two orbits they were 250 miles behind, close enough to get a radar lock-on. They were closing slowly but steadily, in the same plane as Gemini 7 but 15 miles below it, catching up at a comfortable rate. When the target was 30° above them, Wally thrust toward it, transferring Gemini 6 into an intersecting orbit. Two small course corrections kept Gemini 7 centered in their sights, and when they reached a range of 1,000 yards Wally began to decrease his speed according to a carefully calculated schedule.

On the chart Stafford held in his lap, range was R and the closing velocity or range rate was $\dot{R}$ (the time derivative of R). The idea was to keep R and $\dot{R}$ in balanced harmony as each approached zero. If $\dot{R}$ was not decreased quickly enough, Schirra and Stafford would whiz by Borman and Lovell. If $\dot{R}$ was decreased prematurely, orbital mechanics took over and the two Geminis would make a little pirouette in the sky, and the rendezvous process would have to begin all over again. When R and $\dot{R}$ both reached zero simultaneously, the two Geminis would be formation flying, at the same speed in the same orbit, and orbital mechanics could then be ignored for the first time.

With Stafford calling off numbers and Schirra firing the forward firing thrusters to slow down, Gemini 6 coasted up to Gemini 7 and stopped just a few yards from it, the world's first space rendezvous and a damned smooth one at that. Wally had used only 38% of his available fuel.

While in formation, each crew looked at the other's spacecraft. They were a glorious sight—the charcoal gray of the entry module, the adapter section glistening hospital white in the sunshine, and a thin gold thermal shield at the rear of the adapter crinkling with every motion. For Frank and Jim, a minor mystery was solved. From time to time they had noticed a strange shadow pass over the nose of their craft. They knew they were alone and above all clouds, but still . . . it was eerie. Then Schirra informed them that they were trailing some straps covering an explosive cord that severed connections between them and the Titan II. It was those straps, swinging lazily in the sunlight, that were casting the mysterious shadows.

After Gemini 6 departed, Borman and Lovell squirmed and counted the hours. They had been up 11 days and were not looking forward to 3 more. For comfort's sake, they had taken off their special lightweight pressure suits and rode in their underwear with the suits stuffed behind them, leaving them even more cramped than ever. Jim Lovell found his thoughts going to his legs and the fact they were useless in weightlessness. They didn't have to hold him up and they couldn't move him about. They were just *there*, taking up valuable space and contributing nothing. Maybe whoever had recommended a legless Mercury pilot to Gilruth had been right after all. Borman fretted about the condition of the spacecraft and wondered if it would make the full 2 weeks. He had trouble with a couple of thrusters and the fuel cells were slowly dying.

In such a tiny space even the simplest things become complicated. Finding the right meal in the right storage box, finding the scissors to open the package, losing some of the individual containers as they floated about, filling up some of them with water, finding a place to dispose of the waste containers, which seemed—even though empty—to have grown in size—such housekeeping chores were a damned nuisance. Urination and defecation, sleeping, eating, working— everything that had its own place and pace on earth, aboard Gemini 7 happened right *there*, in the awkward curve of that damned ejection seat. At 5'11", Lovell was too tall to completely straighten his body; either his feet hit the bulkhead under the instrument panel or his head bumped into the hatch, or both. If he extended his left arm out to the side, it hit Borman in the face. He couldn't move his right arm to the right at all because he was wedged against the right-hand bulkhead. Neither he nor Borman could change underwear or remove the biomedical sensors taped to their torsos. Ah, the life of an astronaut! If Gemini astronauts in general were twins, certainly these two were Siamese twins, joined together for 14 days of bizarre intimacy.

Today Jim Lovell remembers those 2 weeks fondly. Sure there were minor annoyances, such as losing a toothbrush and having to share the remaining one with Borman, and 14 days can seem an eternity, but he never really got bored looking at the earth. "Where else," he asked me, "can you begin lunch over

Baja, California, and finish over Africa?'' Surrounded by computers and fancy timepieces, he and Frank kept track of the days by scribbling on the spacecraft wall: ++++ ++++ ////.

They were a remarkably durable pair. Back on the carrier deck they looked white as ghosts, exhausted and unsteady, but their physiological measurements were much better than expected, better in fact than those of the Gemini 5 crew had been after 8 days. Like all Gemini crews, Frank Borman and Jim Lovell were awarded the NASA Exceptional Service Medal, but they deserved more— much more.

Chuck Mathews looked back on 1965 as a year of gargantuan achievement for his diminutive spacecraft: five manned flights, all successful, and nearly all the program objectives accomplished. Only docking—the physical locking of Gemini to Agena—remained to be demonstrated, and he had five more flights in which to do that and, more important, to train flight and ground crews in some of the intricacies of rendezvous: rendezvous in daylight or at night, rendezvous from above or below, rendezvous at fast or slow overtaking speeds. These variations, and more, might be encountered around the moon.

Gemini 8 was the first docking flight and it turned out to be an exciting ride. Neil Armstrong, the commander, was the most experienced test pilot in our group of 30 astronauts, and Dave Scott was a former Air Force fighter pilot with a recent master's degree from the Massachusetts Institute of Technology. Their Agena got off without blowing up, and Gemini 8 had a routine ascent and rendezvous. Docking, as we pilots suspected, turned out to be a piece of cake, like mid-air refueling without any wing wash or turbulence.

Seven hours after liftoff Gemini 8 was comfortably attached to the torpedo-like Agena when the two spacecraft began to roll to the left. Armstrong and Scott had been warned that the Agena's control system might be acting up, so they turned it off, but the roll continued. They decided to separate from the Agena and as soon as they did, it became clear that the problem was in the Gemini's control system, not the Agena's, because Gemini 8 began to gyrate wildly.

''We have serious problems here . . . we're tumbling end over end,'' Scott reported to the ground. By this time they were whizzing around at the rate of one turn per second, enough to make them dizzy and cause their vision to begin to blur. Of course, while this was going on they were working as fast as they could to isolate the problem. It was clear by now that a thruster was firing continuously, but which one?

There were 16 thrusters and it was not possible to isolate the guilty one, at least not before they got to spinning so fast they would no longer be able to see their switches. In desperation they turned off power to all 16 and activated a separate group of thrusters, the reentry control system. Within seconds they were able to gain control. Then by turning on the 16 thrusters one by one they were able to identify the short-circuited culprit, but by that time it didn't really matter, because it was a firm rule that once the reentry control system was activated they had to return to earth without delay.

At Houston people in Mission Control were scurrying around like ants whose hill had just been stepped on. There were three main factors to consider: the orbital path over the ground, the weather around the world, and the deployment of suitable recovery ships. Flight Director Gene Kranz made a quick decision, and as Gemini 8 passed over the Congo the crew fired the retrorockets and started a gradual descent into the Pacific Ocean south of Japan, where a destroyer picked them up, safe and sound after their 10-hour trip, happy to be alive but frustrated about all the good things they had planned to do later in the flight— especially Dave Scott, who had trained long and hard for EVA.

With Gemini 9 the problem switched back to the target vehicle again. The Atlas took a nose dive and buried their Agena deep in the Atlantic Ocean. The Gemini 9 commander, Tom Stafford, had been through this once before on Gemini 6, so he and his affable copilot Gene Cernan shrugged and waited for a better day. Instead of a second Agena, however, they got something called the augmented target docking adapter—the front end of an Agena without its motor and tanks. They could practice docking with it but could not use an Agena's power to change orbits. Unfortunately, when Tom and Gene completed their rendezvous with the adapter, they discovered that the shroud covering its nose had not released, so they were not even able to dock. To Stafford the Agena looked like an "angry alligator."

Unlike Dave Scott, Gene Cernan got an opportunity to go EVA. His equipment included a huge backpack called the astronaut maneuvering unit (AMU). It had two control arms that swung down on either side. Each one contained a hand controller, and, using them to steer, Gene could zip around the sky in true Buck Rogers style. The only hitch was that the AMU was almost as large as the Gemini's cockpit, so it had to be stowed outside in the rear of the adapter section. Tom and Gene dumped cabin pressure and opened the hatch, and then Gene, using a chest pack and umbilical for oxygen, edged his way along handrails back to the adapter.

Here began the world's first weightless wrestling match: Cernan the Sky King versus the Evil AMU. First off the lights weren't working properly and Cernan was working in semi-darkness. But the main problem was the damned instability caused by weightlessness. Everything Gene pushed against, pushed back with equal force, and he tended to go flying off in some unwanted direction. He was exerting nearly all his strength just maintaining his position instead of getting a hammerlock on the AMU.

As he worked, banging around, rigging lights, opening and checking nitrogen and oxygen valves, pulling side arms down into position, activating the AMU electrical power system, attaching umbilical cords, arranging harnesses, etc., Cernan began to sweat profusely. His heart was banging away at between 140 and 180 beats per minute, and it was later calculated that he was burning up

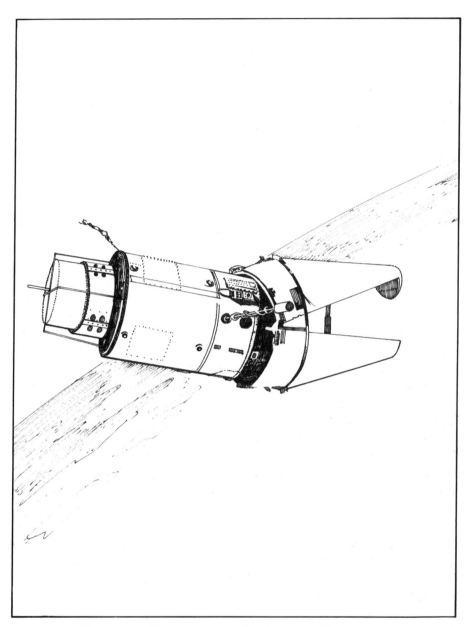

The "angry alligator"—the augmented target docking adapter

energy at the rate of 3,000 BTUs per hour—about like attacking a rock pile with a pickaxe.

Under these conditions Gene simply overpowered the cooling capacity of his chest pack, and his visor began fogging. Resting for a while he felt better but still couldn't see adequately. He decided to give up the good fight and, half-blind, struggled back to the cockpit and closed the hatch with some difficulty. AMU one, Cernan zero.

But it wasn't Cernan's fault and it wasn't the fault of the AMU. The problem was that no one had adequately anticipated that maintaining body position would be such a struggle. If two arms were needed for a task, then two stable foot restraints were required for anchor points so that the body could be pushed and pulled and twisted without floating off somewhere. As badly as he wanted to try out the AMU, it was probably the wisest decision of Gene Cernan's life to stop before he was totally blind and exhausted.

Gemini 10 was next and it was *my* flight, although John Young probably thought it was *his*. He was the commander and I was the copilot. We had the nicest flight plan of the series, I figured, because in 3 days we were going to do a little bit of everything. After what we hoped would be a standard launch, we would overtake our Agena using our own on-board navigation rather than relying on ground calculations. After docking with it, we would use its big engine—a first—to boost us up to a higher altitude so that we could find a second Agena—the one left up there by Armstrong and Scott. In the process we would set a world altitude record. The rendezvous with the old Agena would be done without radar—another first. And, best of all, I was scheduled for two periods of EVA, one just to stand up in the hatch and conduct experiments, but on the second I would use a hand-held maneuvering device to fly over to the Gemini 8 Agena and retrieve an experiment package from it. Chuck Mathews called it "the most ambitious of all the Gemini missions to date." I called it the best.

Another nice thing about Gemini 10 was that launch was scheduled for 5:20 P.M. . Instead of getting up at 4:00 A.M., like most crews, John and I learned—without difficulty—to sack in until noon. True, we were working until 4:00 A.M., but still it seemed a more civilized schedule. I had a little black notebook in which I entered "open items," small things that bothered me or that needed attention. By launch day I had entered 138 of them—and crossed them out one by one. I was as ready as I was ever going to be.

July 18, 1966
The little old ladies of Worcester have been good to me, and my hand-tailored pressure suit fits like a glove. I slip into it, lock helmet and gloves in place, and begin breathing 100% oxygen, to start the process of purging nitrogen from my system. In a sense this is the moment when an astronaut leaves the earth, because when his visor clicks shut, he loses his normal senses of smell, touch,

and even hearing. Inside that impermeable cocoon he feels more like an appendage to a flying machine than attached to the soil of this planet.

Carrying a briefcase full of oxygen, John and I ride in a small van out to the launch pad, and enter the grillwork cage of an orange elevator that clatters up the side of the gantry. On top, people help shove us down inside our Gemini, attach parachute harness, oxygen, and radio connections, reverently close the hatches, and then we are left alone. But we are not lonesome, for over the radio comes a continual chatter from people in the nearby blockhouse and from Mission Control in Houston. It is mostly technical stuff, having to do with the health and happiness of Gemini and Titan, but includes the welcome news that our Agena has made it safely into orbit. Bless that gasbag of an Atlas; I wouldn't want to ride one but I sure appreciate it as a delivery system.

Now it is our turn, and down below our Titan stirs as its two engines swivel back and forth. We feel a little shudder in the cockpit and then all is still again. Finally comes the 10-second countdown, and clutching the ejection ring between my legs in a death grip, I hold my breath and wait for ignition. There it is! Noise, plenty of it, but we feel this machine rather than hear it. A little bump and I know the tie-down bolts have sheared and we are on our way. We jerk back and forth a little as the engines 100 feet below us work to keep us balanced up here, despite gusty winds and sloshing fuel tanks. I can see a tiny patch of blue sky out my window but there is no sensation of speed until a thin layer of cloud approaches. Then—pow—we burst through it faster than my eye can follow. We really *are* moving out. I begin to feel heavier and heavier, and now starts a mild Pogo, a fore-and-aft vibration that causes the instrument panel to go slightly out of focus. At 50 seconds we pass the altitude at which we may use our ejection seats and I let go of the ring, my fingers tingling. Noise and vibration increase here as we approach Mach 1, then smooth out abruptly as we reach supersonic speed and thinner air.

In 1 minute and 20 seconds we have accelerated from zero to 1,000 mph and have climbed from sea level to 40,000 feet. As the first stage tanks empty, we are up over 5 Gs.

Staging, like a popping flash bulb, freezes certain images in its aftermath. All at the same instant the G load hits zero, I am flung forward in my straps, and the window changes from black to red to yellow as the evidence of an explosion envelops us. We find out later that the second stage engine has blasted the top of the first stage, causing the first stage oxidizer tank to explode—without damage to us but causing a spectacular light show visible even on the ground. As quickly as chaos has come, so is order restored, as the second stage hums along quietly, and the window darkens again. Now a heavy hand is pushing on my chest with seven times its normal weight. Before I can get too uncomfortable, however, there is another lurch—and there we are, hanging in our harness, weightless at last.

Out the window is the most amazing sight I have ever seen, a glorious panorama of sea and clouds stretching for a thousand miles in a glistening white light. In an instant our dark little cubbyhole has been magically transformed. These are not ejection seats but thrones, facing out on the universe, and we are wealthier than kings. Alas, in 7 minutes we will be in darkness and my work will begin, but what I have seen already makes the months of tedious preparation worthwhile.

Restrained in its harness, my body doesn't seem much different in weightlessness. There is a slight feeling of fullness in my head and my forearms want to float up in front of my face. But I can definitely tell that we are weightless, because there is a small armada floating around the cabin: washers, screws, pieces of potting compound, lint, dirt—the detritus of the manufacturing process, despite McDonnell's best efforts in its clean rooms.

By now we have assured ourselves that Gemini 10, which we know so well from ground tests, is operating flawlessly as a spacecraft, and it is time for me to begin my navigational computations. The key to these is measuring precisely the angle between selected stars and the horizon. It's not as easy as in the simulator, I soon discover. I can find the stars easily enough. They are old friends with mysterious names: Schebir, Hamal, Fomalhaut, and Arcturus. But having found them, my troubles begin when I peer through my sextant. Then the stars are hard to see, especially in the airglow layer just above the indistinct horizon.

The success of our on-board orbit determination scheme depends not only

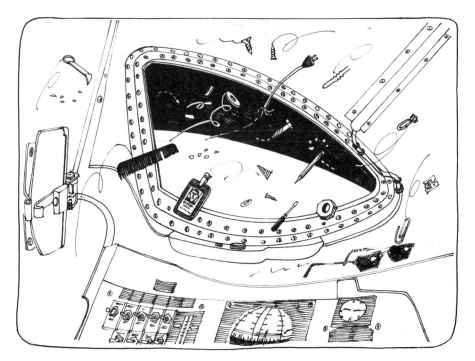

Weightless detritus

on accurate measurements but on tight timing as well, and I find myself slipping farther and farther behind as I fill charts and graphs with hastily scribbled numbers. I compare the maneuvers I have calculated that will cause us to overtake our Agena with the ground's solution. The results are quite different and we are forced to use the ground data, much to the chagrin of Magellan, as John has taken to calling me.

Within 4 hours we have made three maneuvers, or "burns," and are 15 miles below our Agena and closing nicely. From now on we are on our own and the ground can no longer help us. At a range of 40 miles when the Agena is 30° above us, we thrust toward it, to establish an intersecting trajectory. Up until this point we have seen it only as a flashing light, but now we can begin to make out its cylindrical shape.

John is flying and I am alternating between peering out the window and reading numbers from our computer. At a mile out, we are closing at 25 mph, which is about right. But then things start to go wrong. We have gotten off to one side somehow and our closing rate has dribbled off to zero. We have to thrust toward the Agena again and now we are swinging around it in a tightening arc. We have done this before in the simulator and we don't like it a bit. "Whoa, whoa, whoa, you bum!" John yells. We call this curlicue maneuver a whifferdill, and it's the biggest fuel waster in the book. When John finally pulls up next to the Agena we have only 36% of our fuel remaining instead of the 60% we expected at this point. It's a gloomy cockpit.

Our spirits are restored somewhat by the docking. John guides our snout easily and gracefully into the Agena's docking collar, latches snap into place, and a motor in the Agena pulls us tightly together until we are "rigidized," in NASA language.

The Agena has a small dashboard mounted above its docking collar and if we look carefully we can make out lights and dials indicating its state of health. Everything looks okay as we prepare to light its engine to start a second chase to find the Agena left over from the Gemini 8 mission.

The next rendezvous will be trickier because the Gemini 8 Agena, some 100 miles above us, has long since run out of electrical power. Therefore, it has no lights and—much more important—its transponder is dead and cannot respond to our radar's inquiring signal. We will have to make our moves using only our eyes and our platform to measure the angular relationship between Gemini and Agena. I will have to estimate range and range rate by using a sextant to measure how large the Agena appears in my field of view and, keeping one eye on my clock, how rapidly it is growing.

Our own Agena will get us there and then we will discard it, like a railroad switch engine that has finished its job. I communicate with an encoder, a lever with which I send the Agena three-digit messages. For example, 251 means Please turn on your lights. When we are ready to ignite its big engine, all 16,000 pounds of thrust, I tell it 041-571-450-521-501 and then hold my breath. An

automatic sequence has begun. At the instant of ignition I am disappointed to
see only a string of snowballs shooting out the back of the Agena in a widening
cone. They are beautiful against the black sky, but I am about to tell John we
have a malfunction when—wham!—the whole sky turns orange-white and we
are plastered against our shoulder straps. It kicks like a mule, this engine, and
in just 14 seconds it has added enough energy to our orbit that we will swing
up to 475 miles—a world altitude record.

We are over Hawaii, so our apogee will occur 180° from now, on the opposite
side of the earth, over the South Atlantic. As it happens, this region is where
the inner Van Allen radiation belt dips earthward and we will be skimming
the bottom of it. In the meantime, the big engine subsides, coughing and belch-
ing, and in our wake we are treated to a glorious Fourth of July spectacle.
There is a golden halo surrounding the entire Agena. In its center, like a
Roman candle, the engine is spitting out residual chunks of fuel and oxidizer,
some tiny as fireflies, others large as luminous basketballs. The sun directly
behind us spotlights these glistening silver spheres as they silently fade into the
black sky.

Once again our attention returns to the cockpit. In spaceflight it does not pay
to linger over past events because there is always something just ahead that
requires full attention. But John can't help giving one last tip of his hat to the
rambunctious Agena: "When that baby lights, there's no doubt about it."

Our next concern is the amount of radiation penetrating through the walls
of the spacecraft into our bodies. There's not a hell of a lot we can do about
it right this minute because we are going up like a runaway elevator, but if
our dosimeters read too high we will have to use the Agena again to lower our
apogee when we get back over Hawaii. However, the dosimeter readings are
unbelievably low, only one-tenth of the predicted values, so we can settle down
for the night. It is 2 o'clock in the morning, Cape Canaveral time, and we have
had a hell of a day. On top of everything else, my left knee has been aching
for the past couple of hours, an indication that my preflight nitrogen purge has
not worked 100% and I have a mild case of the bends. John is morose because
of our excessive fuel usage; I am hurting despite a couple of aspirin. We are
not in a happy mood after our first day in space.

In space the terrestrial concept of night and day has to be modified somewhat.
Day means we are in view of the sun, night that we are in a planet's shadow.
On earth we know that (except for the strange lighting conditions near the North
and South poles) our 24-hour cycle will consist of one period of light and one
of darkness. Not so in space. A spacecraft in low orbit will circle the earth
16 times in those 24 hours. Each orbit in turn consists of one period of sunlight
and one of darkness. During a terrestrial 24-hour "day" an astronaut will see
16 dawns and 16 dusks. He simply ignores the lighting conditions out his window
and believes his watch, which he keeps on Cape Canaveral or Houston time.
His own body clock is still tuned to earth and will cause him to feel awake

or sleepy just as if he were still on the surface at that location to which his circadian rhythm has become adjusted.

Going to the moon is a different story. Between earth and moon there is no "night" in the terrestrial sense because there is no planet to hide behind. During the 80 hours or so of perpetual sunlight between the two bodies the astronaut again ignores what he sees out the window, and keeps his watch—and his body—on Houston time.

Back in the Gemini 10 cockpit our method of turning out the lights is to clamp thin metal plates over the two windows. When we unpack them we discover that some kind soul had pasted on each a photograph of a wildly beautiful girl. These two intruders seem absurdly out of place in our mechanical cubbyhole, but up on the windows they go, and we are left in the darkness to doze fitfully.

The first night in space is never a restful one. Too much has taken place on launch day, the new environment is too strange, and it's asking too much for the body to unwind. In addition, my knee hurts, I have flunked my Magellan test, and our excessive fuel consumption may cause cancellation of one or both of tomorrow's EVAs. On top of all that, my hands in weightlessness have assumed a life of their own and dangle at a ridiculous angle out in front of my face, like the forearms of a praying mantis. I am afraid if I go to sleep with them in this position they may bump into some important switch. Also my head needs a pillow. The rest of my body is content to hang there, in constant free-fall, but my head and neck want to feel some reassuring pressure on them. Finally I turn away from John and wedge my head as far as I can into the upper right corner of the cockpit, between the ejection seat and a storage locker. After a long time I doze off, for perhaps 2 hours of solid sleep. Then it's time to get going on Day Two.

Things seem better today. My knee doesn't hurt anymore and our radiation counter reads .78 rad, which is an insignificant dose. We ignite the Agena engine again to make our orbit come closer to that of our silent target, and I prepare for my first EVA. This first peek outside will take me no farther than a yard above my seated position, but I am apprehensive nonetheless. I get apprehensive any time the cabin pressure goes down to zero. Not that I don't want to do it. On the contrary, for months I have daydreamed about seeing the world beyond the narrow picture frame of my tiny Gemini window.

I open the hatch at sundown and it swings freely in my hand. That is a good omen because on previous flights the hatch has been hard to open and even harder to close. It is pitch black as I cautiously emerge, waist high in the open hatch. I have attached extensions to my oxygen hoses and radio leads to allow me to do this. My first task is to measure the ultraviolet signatures of some Southern Constellation stars by photographing them using special film, 20 seconds per exposure. It is relaxing work and once I get the hang of manipulating the camera in my heavy pressure suit gloves, I begin to look around.

My God, the stars are everywhere, even below me. They are somewhat brighter than on earth, but the main difference is that they don't twinkle. That effect, which the astronomers call scintillation, is caused by the wavering of the atmosphere, scattering the starlight as it passes through. Up here, above the atmosphere, the stars appear rock steady. The planet Venus also appears absurdly bright, like a 50-watt bulb in the sky. I have to convince myself it *is* Venus, and not some UFO, by confirming its position among the stars. Down below, the earth is barely discernible. The moon is down below the horizon and the only noticeable light comes from lightning flashes along a row of thunderstorms on earth. An occasional meteor hurtles into the atmosphere, and it is a strange sensation to look *down* on its fiery trail.

The sun comes up with a fierce burst of piercing white light and shortly thereafter my eyes begin to water. My sun visor is down and I have lowered my head, turtle-like, into my suit to keep my eyes in the shade, but I still can't see well enough to change the f-stop on my camera. I pass the camera back in to John and ask his help. Now he tells me he is having similar problems. When I repeat that my eyes are still watering, he says, "Mine are, too, Mike. I can't see a gosh-darn thing." He sounds cheerful about the whole thing, but then adds: "Okay, come on back in. Let's close the hatch."

I have practiced this maneuver a hundred times, stuffing my huge pressurized suit into the tiny cockpit. It has to be done in a precise sequence: the first and key step is to jam my legs down under the instrument panel until my knees are 6 inches below the rim of the panel. I can then use my knees as a fulcrum, bending them to force my stiff upper torso to rotate down below the level of the hatch. If I have done this maneuver properly, I can reach up and swing the hatch inward until a latch clicks, and then all I have to do is crank a handle to lock it. If I have not gotten my legs far enough down, the hatch will bang against the top of my helmet without closing, and I will have to pop out and repeat the whole process. It's a tight squeeze, but this time my adrenaline pump is working just fine and when I swing the hatch down I am rewarded with a reassuring clunk as the latch engages. Within a couple of minutes we have filled the cabin with oxygen and we compare notes with Mission Control. None of us can figure out what has caused our eyes to become irritated.

We get on with our work, this time an experiment to detect the precise location of the murky horizon, using a photometer to track a star and measure the decreasing intensity of its light as it passes through the atmosphere and disappears from view. Our eyes gradually return to normal except for some red puffiness around them, and I realize that Day Two hasn't been half bad. Thanks to our Agena, whose fuel we have been using for maneuvers, we still have 30% of our own propellant. Ground Control tells us they are devising a test to try to pinpoint the cause of our eye irritation. We eat a good meal and hit the sack. Tonight this little cockpit doesn't seem such an outlandish place to sleep. Besides, I am bone tired, so I drift off quickly with a feeling of optimism about tomorrow.

The next thing I know 7 hours have passed and there is a voice in my ear. It's Houston, calling us to get started on Day Three. While we are eating breakfast they outline a test to determine if our eye irritation has been caused by our carbon dioxide absorbent, lithium hydroxide, somehow escaping from its container. During my next EVA I will be operating on a separate oxygen supply fed through a 50-foot umbilical, so I am less concerned about myself than John, who will be hooked up to the same oxygen hoses as yesterday.

I obviously need to see without impediment in order to go over to the Agena and retrieve an experiment package from it. But so must John also be able to see clearly, to fly the Gemini up next to the Agena and keep it there, in precise position while I am scrambling around outside. He will also be essential as a pair of eyes guarding my rear and letting me know if my trailing umbilical cord is in danger of wrapping around the Agena, or getting snagged on it. We pass the lithium hydroxide test with flying colors and that is a relief.

Next we make two small orbit adjustments using our Agena's power and then we separate from it. The Agena is an old friend by now and we hate to see it go, but we also enjoy being free—for the past 40 hours our view has been impeded by the Agena stuck on our nose, and we have gone around the world craning our necks like someone in a caboose trying to see around the locomotive up front.

We are only 8 miles below the Gemini 8 Agena and closing at a slower, more cautious rate than we used to catch our own Agena, in deference to the fact that we have no radar to assist us. We see it for the first time, a tiny speck at which John points our nose. While he tracks it precisely I measure the rate at which our nose's angle above the horizon is increasing. By comparing these actual angles with a chart full of theoretical ones, I am able to calculate when we should depart this orbit and transfer to an intersecting trajectory.

Getting there without wasting fuel is only half the problem; the other half is getting there on time—before sundown, because with no radar to tell us range and range rate, we will surely slide by it in the darkness. This entire scheme has been calculated months before, to make the best use of our 55-minute ''day,'' but it's a tight schedule nonetheless, one that calls for us to reach the Agena just 5 minutes before sundown.

Things look good as we close in. The Agena grows from a dot to a cylinder and with my sextant I can now measure the angle it subtends. By comparing its growth with my clock, I give John estimates of our range and closing speed, but beyond that I can only shout words of encouragement and antiwhifferdill sentiments. John brakes to a halt, smoothly this time, and now we are riding serenely next to the Agena, with our fuel gage reading 15%. Not bad!

Normally we would like to rest a bit at this point, but we cannot. The sun goes down, abruptly as usual, and John and I each pursue our own demanding tasks. John flips on our searchlight and for the next 37 minutes of darkness flies formation with the Agena. This is hard work because he must never allow it to wander out of the tiny circle of illumination from the searchlight, but at the same time he must be parsimonious with our fuel, making as few corrections as possible.

While John is doing all this I am busy as hell unstowing and installing all the gear I will need for EVA. There is a 70-step checklist I must go through, and these are crucial steps because once we dump cabin pressure I am utterly dependent on this equipment I am assembling now for the first time.

Maneuvering unit

The three main items are my chest pack, umbilical cord, and maneuvering unit. Then there are cameras, experiment packages, checklists, and various minor pieces of hardware. It is not a tidy process, due mainly to the bulk of the chest pack and umbilical cord. The latter has been tightly coiled under John's feet, but when unstrapped it expands, all 50 feet of it, and tries to fill the whole cockpit, or at least my side of it.

The function of the umbilical cord is to feed high pressure oxygen from the spacecraft to the chest pack, where a regulator reduces the pressure to 3.7 psi and feeds it into my suit. The cord also contains a nitrogen hose, radio wires, and a stout nylon tether to prevent me from pulling something loose. The nitrogen is routed from a spacecraft tank through a connector just aft of the cockpit to my maneuvering unit, which is a simple, gun-like, hand-held device that squirts a jet of nitrogen out through two nozzles pointed aft and one pointed forward. It has a two-position trigger that I squeeze between my right thumb and index finger. By pointing the gun in precisely the right direction and squeezing the appropriate trigger, I can, in theory, maneuver myself in any desired direction.

As dawn arrives I am on schedule and ready to go. As messy as all the equipment is, with the bulky pack strapped to my chest and the umbilical floating everywhere, I think I have a better arrangement for EVA than Gene Cernan did. At least I have been able to assemble all the gear in the pressurized safety of our cabin, instead of back behind in the adapter section, making critical connections in a vacuum.

John pulls a ring to dump cabin pressure. It takes a little while for all the oxygen to escape, like water draining from a bathtub, and while I am waiting I review some of the things I must do and not do. High on the list is a requirement to coordinate with John the firing of our thrusters. At one time or another he will use all 16 of them, to keep us in proper position next to the Agena, but if I am next to one he must wait because its hot exhaust gases may blow a hole in my suit or gloves. Each thruster is a small rocket motor, and dangerous. Thruster No. 16 is of particular concern because it is located next to the connector for my nitrogen line and equally close to a micrometeorite experiment that I am to retrieve and return to the cockpit.

When the time comes, again the hatch opens easily and I move out of it gingerly. First I raise a handrail that has been flush against the side of the adapter and using it, I move to the rear and remove the micrometeorite detection plate that has been exposed the past 2 days. While back there I caution John not to fire No. 16, the one that he needs to cause the Gemini to translate downward. "Okay," he replies, but then adds, "Well, babe, if I don't translate down soon, we're going to run into that buzzard." The more nervous John gets, the more he calls me "babe." By now I am clear of No. 16, I hand the plate in to John, and return to plug in my nitrogen line. Again I warn John and again I get the same answer, "Okay, we're going to hit this thing if I can't translate down pretty soon." I get out of the way and John maneuvers.

Then I move back and, with some difficulty, connect the nitrogen line. The problem is twofold: first, with nothing to restrain my body, my legs are flailing back and forth in response to the slightest torque that my arms put on the rail or the connector. Second, I have missed on my first attempt to stab the connector with the fitting on the end of my umbilical. The sleeve on the fitting has sprung forward and must be recocked, but that is a two-hand operation, leaving me no hands to hang onto the rail. I let go for an instant, recock the sleeve, and grab the rail again. In the process I swing wildly and my legs bang up against the side of the spacecraft. John feels the commotion and so does the Gemini's attitude control system, which reacts to this unwanted motion by firing thrusters to restore equilibrium. "Boy, Mike, those thrusters are really firing. Take it easy back there, right?"

I return to the cockpit and get ready to make the great leap for the Agena. I realize that I have been oblivious to everything around me and to the fact that we are traveling nearly 18,000 mph. Actually the speed doesn't matter, as long as our speed relative to the Agena is zero. Nor does it matter that we are 250 miles up; I am vaguely aware of the earth slipping by below, but it is not in this equation. All that matters is us and the Agena.

I talk John into position because I can see much better than he can, peering through his window. When we are 6 or 8 feet below and behind one end of the Agena, I ask him to back away a little and then I tell him I'm ready to jump. "Take it easy, babe," he replies.

Actually, jump is the wrong word. Gently I push away from the open cockpit, trying to shove equally with each hand so that I will come out straight and not like a corkscrew. As I float up and slightly forward, I note with relief that I'm not snagged on anything and I seem to be headed in the right direction with no tendency to pitch, yaw, or roll. Within a few seconds I collide with my target, the docking adapter on the end of the Agena.

The experiment package I am to retrieve is on this end, but still it's a bad place to land because the docking cone has a smooth, tapered lip that is hard to grasp with my inflated pressure suit gloves. When I grab it I realize I need to move to my left to get the package and I head that way, going hand over hand. As I move I dislodge part of the docking apparatus, an electric discharge ring that springs loose, dangling from one attach point. It looks like a thin scythe with a wicked hook. I suspect it is fragile and will pull loose easily, still it is made of metal and I don't want to risk getting me or my umbilical ensnarled with it. Now I have reached the package. I try to stop but the momentum in my lower body causes me to keep going and peels my hands right off the Agena! First my right, then my left slip free and now I am turning lazy cartwheels somewhere above and to the left of everything that matters.

All I can see is black sky. While I am trying to figure out what to do next, the Gemini swings into view down below me. "Where are you, Mike?" John doesn't sound too happy.

Michael Collins on EVA

"I'm up above. You don't want to sweat it. Only don't go any closer if you can help it, okay?" My concern is that now I can see both Gemini and Agena, and I notice that a loop of my umbilical is over by the docking collar end of the Agena. I am afraid that if John gets any closer to the Agena, which he may not be able to see right now, my umbilical will get tangled up with the dangling electric discharge ring. No sooner has this problem arisen than it disappears, to be replaced by another.

I have swung out to the end of my 50-foot umbilical now, and am moving away from the Agena and *tangent* to the Gemini. This poses a new danger because the laws of physics tell me that if I do nothing I will wind around the Gemini like the string of a top. As my radius decreases, conservation of angular momentum decrees that my velocity will increase and I will splat up against the side of the Gemini at a very high speed. (This is what allows ice-skaters to spin faster as they pull their arms in close to their bodies, but it isn't what I had in mind.)

My salvation is my maneuvering unit, my gun. With it I can zero, or at least decrease, my tangential velocity. I reach for it. It's gone! I grope around and find the hose leading to it, and reel it in, and finally I have the gun in my right hand. By now my arc has taken me high off to one side of the Gemini and I am swinging around toward the rear of it. I start spraying nitrogen like crazy and while I am unable to stop my motion completely, I do slow it down enough to make an airplane-like approach to the cockpit, gliding in from the rear. I have never intended to get back here and I don't like it because this is where all the thrusters are. I tell John, "I'm back behind the cockpit, John, so don't fire any thrusters." "Okay," he says, but adds, "we have to go down if we want to stay with it."

That is the worst possible news. Not only will the thrusters be blasting right at me, but if the Gemini goes down I am going to skim over the top of it and begin another curlicue. "Don't go down right now. *John, do not go down!*" I must sound sufficiently worried because he mutters, "okay." Now things are better. I am past all the thrusters except the one by the nitrogen connection: "John, do not fire that one bad thruster, okay?" "Which bad one?" he wants to know. Damn, it's no time for numbers games. "You know, the one that squirts up." "Oh," he says nonchalantly, "16." My approach to the open hatch certainly is not graceful but will do. I bang into the hatch faster than I would like but manage to hang on. I get back inside part way and reel the umbilical in after me.

This time I'm going to have to do something different. "Okay, John, want to give it a new try over there?" "Yes." "Okay, let's try it one more time." "Okay." John maneuvers us into position again but this time a bit farther away from the Agena. It looks as if the docking cone is 15 or 20 feet in front and above me when I leave the cockpit. This time instead of pushing off, I use the gun, pointing it at the end of the Agena.

Up I glide, slowly—almost miraculously, it seems to me—but my left boot snags on something (the top of the instrument panel?) and causes me to start pitching face down. If this continues I will hit the Agena with the small of my back so I start squirting away with the gun to pitch back up. In the process the gun not only rotates me but raises me as well, and I discover to my horror that I am about to cruise over the top of the Agena. I manage to make one last frantic correction and as I go by am able—just barely—to snag the Agena with my left arm. As my body swings around this pivot point I let go of the gun and plunge my right hand down inside a recess between the docking cone and the main body of the Agena. I've got a fistful of wires now and I'm not about to get thrown off.

After all this I have lost my bearings and don't know which way to go to find the experiment package. I am concerned about the dangling hook of the electrical discharge ring and so is John. "See that you don't get tangled up in that fouled thing," he admonishes me. "Yes, I see it coming." I continue, hand over hand, past it. "Don't get tangled up in that thing. It's going on behind you now." By that I guess he means it is flapping back and forth in response to the disturbing torque of my flailing body. I can't worry about it now: "If it starts to look bad, let me know. I'm going to press on up here."

Finally I make it around to the experiment package, which is protected by a fairing and held in place on rails. I depress two release buttons and give a jerk. The fairing springs loose and attached to it by wires is the package. Success! And not a minute too soon. The Agena, which started out nice and stable, is now gyrating around in response to all my banging into it. John is having a hard time staying in position, and he's still worried about my umbilical getting fouled up somehow. "Come back," he warns, "get out of all that garbage . . . just come on back, babe." That suits me fine. "Don't worry. Don't worry. Here I come. Just go easy." This time with no tangential velocity to worry about, I am coming back the easy way, hand over hand on my umbilical— but *slowly*, to avoid splatting up against the side of the Gemini.

Back in the cockpit I discover to my sorrow that my camera is missing. A 70mm wide-angle Hasselblad, it had been attached by a bracket to my chest pack, but somehow worked its way loose and floated off. For years it will orbit by itself, full of spectacular pictures of Gemini and Agena. But there's no time to brood about the loss, as we have other things to worry us. John can no longer see the Agena.

"We're not going to hit the thing, are we?" he asks. "No, we're clear. I'm watching it. We're good and clear." The Agena is free to return to its status of drifting derelict, of no further use to us or anyone else. I make a report to Houston on the EVA and the difficulties I have encountered, and they order me back inside because we have reached the red line on our maneuvering fuel.

I disconnect the nitrogen line and stuff the umbilical down in the cockpit as best I can. The damned thing is wrapped around me a couple of times and prevents me from forcing my knees down around the instrument panel. My visibility downward is severely limited by my helmet and I can't see exactly what the impediment is. Bad news! I really don't want to go back outside and unravel myself. Fortunately John can see my lower body, and one advantage of the tiny cockpit is that he is close enough to reach over with his right arm and start yanking on the umbilical.

I back part way out and now he has freed me except for one persistent last loop around my legs. No problem. My legs slide down easily and I use my knees around the lip of the instrument panel to torque my body down far enough to swing the hatch shut. Umbilical is everywhere, a nearly opaque mass of white coils that obscures the instrument panel and John and everything else.

While I am cranking the hatch locking handle, Houston asks John for a propellant quantity reading. "Get serious," he growls. Fifty feet of thick umbilical has filled the cabin like an ice fog. I thrash around as best I can, trying to gather the umbilical into a manageable package, and in the process manage to bump one of the radio switches so that we lose contact with Houston. Strange that they are so silent.

When we discover my error, John tries to be funny: "He's down in the seat because there is about 30 feet of hose wrapped around him. We may have difficulty getting him out." I try to be funny: "This place makes the snake house at the zoo look like a Sunday school picnic." It's pretty thin humor, but it's our way of saying we are most grateful to be back inside, all in one piece, with a comfortable 5 psi of oxygen all around us.

We have one more short EVA, a simple hatch opening to allow us to jettison equipment we no longer need. We prepare a huge duffel bag into which we cram the umbilical, chest pack, gun, empty food packages, and everything else we no longer need. Our original stowage list contained 159 items, and it will be a relief to get rid of some of them. The third hatch opening is routine, but to be on the safe side I wedge myself way, way down before dumping cabin pressure. I fling the bag, slam the hatch shut with a good 6 inches to spare above my helmet and we pump back up to 5 psi. With any luck at all we have trusted the ladies of Worcester and their glue pots for the last time.

With only 7% propellant remaining, the rest of the flight is relatively quiet. We use most of our 7% to lower our perigee from 250 to 180 miles, a precaution in case one of our four solid retrorockets refuses to fire tomorrow. We spend the next couple of hours on experiments and then—for the first time in 53 hours—we have some free time. Actually we have a 10-hour sleep period, but we certainly don't need that much sleep. With no Agena stuck on our nose, and running low on fuel, we let the Gemini drift and wander.

In a fighter plane I have done loops and spins, but I have never flown sideways or backwards. Now we are doing that and more, tumbling slowly, smoothly, aimlessly. As we turn, the blunt snout of our spacecraft traces graceful arcs in the sky, sometimes in front of our direction of travel, sometimes to one side or behind. It is a stately roller coaster ride, utterly silent, with no jerking around and no hollow feeling in the pit of the stomach. It is fascinating. If this is what the old Edwards test pilots meant about an astronaut just being along for the ride, being a canned man, then I'll have to tell them they don't know what they're missing.

The view out the window is worth all the months cooped inside a simulator and trapped on my back in a clean room. At 200 miles above a sphere whose radius is 4,000 we are just skimming along one twentieth of a radius above its surface. The atmosphere itself is thinner than the rind of an orange and we are just barely above it. The curvature of the earth is readily apparent but it is not the dominant impression. Nor is the speed the blinding blur of a race car. Our higher velocity is balanced by our altitude so that the ground goes by the window at about the same rate as in an airliner.

The colors below are unchanged except that the unfiltered sunlight makes clouds and water brighter and shinier. The sky is absolute black instead of blue. Yet despite the similarities, the feeling I get looking out the window is totally different from anything I have imagined. I am seeing a planet for the first time— not in its entirety, but enough for the mind to capture its totality. Those aren't counties going by, those are continents; not lakes, but oceans! Look at that, we just passed Hawaii and here comes the California coast, visible all the way from Alaska to Mexico. San Diego to Miami in 9 minutes flat! If I miss something, not to worry, it will be back in 90 minutes. Up here there are no misty days, no gloom—all is glorious sunshine. This is a better world than the one down below.

It's time to eat and I am famished, having missed lunch because of the press of other business. I have filled a plastic tube of dehydrated cream of chicken soup (my favorite) with water. Cold water, alas, but that is all we have. After kneading it for a couple of minutes I snip off the end of the feeding tube with my surgical scissors. It is a trivial use of the scissors, which I know from practice are powerful enough to chop through a stout umbilical cord. Anyway, I finally get my first swig of soup and it's the best thing I have ever tasted, finer even than a martini at Sardi's or the pressed duck at the Tour d'Argent. I supplement the soup with bacon cubes and wash them down with grapefruit juice. As we cartwheel along, my stomach gurgles happily and my eyes feast on the spectacle outside the windows.

The Indian Ocean flashes incredible colors of emerald, jade, and opal in the shallows around the Maldive Islands. Beyond is the Burma coast and green jungles, followed by mountains, coastline, and fires burning near Hanoi. Now the sun glints on verdant Formosa, the shape and color of a well-tended gardenia leaf. Just south of the island there are interesting surface ripples, intersecting in a clearly visible pattern indicative of currents that I think must be of interest to fishermen. Then back over the Pacific again and a race toward Hawaii and the California coast. I could stay up here forever! Whoops . . . well, at least until retrofire time, 14 hours from now.

I drift off to sleep almost reluctantly. I want to savor not only the view but the day's accomplishments. The tricky optical rendezvous, of course, but it is the EVA that sticks in my mind. Over and over I grab that Agena and slip off. Over and over I cartwheel through the black sky. Over and over I see good old thruster No. 16 about to explode in my face, and the loose hook on the Agena about to enmesh me. I suppose I should have been scared but it's like high wire artist Philippe Petit said: "I am never afraid on the wire. I am too busy."

Day Four begins with the usual bugling from below—in this case our ground station in the Canary Islands pestering us until we are awake. After breakfast I hook a urine bag to the overboard dump valve and am rewarded with what Wally Schirra calls the constellation "Urion"—a snowstorm of frozen particles. Cascading out in an irregular stream they whiz past the window and tumble off into infinity, glistening virginal white in the sunlight instead of the nasty yellow we know them to be. This is a fairyland up here, far above the forgotten squalor of the world below.

As we make our final orbit before retrofire we thank the folks in the various tracking stations we pass over. "Have a good trip home," Canary says. John makes his speech: "Roger, thank you very much. Enjoyed talking to you. It's been a lot of fun . . . want to thank everybody down there for all the hard work." This is more than just a pro forma courtesy. In our case, at least, because of our excessive fuel usage, the troops in Mission Control have really been put to the test. They have done a masterful job of reshuffling the schedule and getting extra work from our Agena, so that we have accomplished almost everything we set out to do.

Retrofire is a serious business. We prepare for it with a solemnity that exceeds even that of launch. Every item on the checklist is marked off carefully, and then the marks are checked. We even discuss *how* I am going to punch a particular row of buttons. "Push them in the center; push them down hard and hold them down for a good fat one second," John lectures me. I am listening carefully: "Roger." He adds, "Have a little separation between them." "Yes, I will. About 2 seconds?" "No, make it 1." "Okay."

As retrofire time approaches, I find myself peering out the window for the tenth time, to make sure that we really are pointed with our heatshield forward, so that the retrorockets will slow us down and not speed us up. If we mess this up somehow, we will become a permanent earth-orbiting satellite. Now we throw switches that chop the fuel and electrical lines leading aft to our adapter section and then we jettison the adapter, exposing the four blunt solid propellant rocket motors. Canton Island comes on the air with the inevitable countdown . . 3 . . 2 . . 1 . . RETROFIRE!

After 3 days of weightlessness, my body is supersensitive to acceleration.

I feel each of the four rockets, firing one after the other, slapping me in the back. It's only ½G but it feels more like 3. "I count four beautiful ones, John babe." "Yes," he replies, and then reports to Mission Control. "That was a superfine automatic retrofire: 303 aft; 5 right; 119 down." He means that our velocity has been reduced by 303 feet per second in the aft direction and 119 feet per second down toward the earth. This is almost perfect. Now we jettison the retropack, exposing our heatshield, and John concentrates on steering toward the point that our computer thinks our landing ship (the aircraft carrier *Guadalcanal*) is located. It's like making gliding turns in an airplane except that we are upside down and backwards, head down for maximum lift and heatshield forward.

I can tell that the ablative heatshield is doing its job because we are developing a tail. Tenuous at first, then thicker and more startling, it glows an eerie red and yellow. John has seen this once before but I have not. "Man, that's starting to look like something . . . look at that son-of-a-gun burn!" Occasionally a small chunk of heatshield breaks loose and adds sparkle to the halo. As the G load builds to 4, I feel we have it made: "Fly that thing, John! You're doing a beautiful job." The Gs taper off now, and at 38,000 feet we let fly our drogue parachute.

Instead of stabilizing us, we begin to swing back and forth in a wild arc that takes us 25° to either side of our supposedly vertical descent. "Shoot," says John, and deploys the main chute early. It fills our windows with red and white reassurance. Houston reports that they can see us on TV, so we must be coming down in the right place. When we swing down to the horizontal position we can tell that we are rotating on our parachute. The clouds are whizzing by our windows sideways! I figure this strange motion is going to increase our descent rate. "Boy, we're going to hit like a ton of bricks," I announce. Amazingly we don't, but plop gently into the Atlantic. We must have caught the lip of a descending wave.

It is a soft and gentle day and the sea is mercifully quiet. My ears are blocked from the swift descent, and there is an unaccustomed aroma of burned chemicals in the air. Outside my window nose-mounted thrusters hiss and smoke, emitting an occasional tendril of weak flame. The ocean air is moist and fetid and hot, and I realize suddenly how artificially dry and cool we have been in our mechanical cocoon for the past 3 days. The thick EVA protective layers of my pressure suit are a suffocating blanket, and I feel I am filling up the suit with sweat. A helicopter flashes past our prow and then swimmers appear to surround us with a stabilizing rubber collar. Gemini 10 flies no more. We are awkward trespassers here, at the mercy of a new set of EVA experts clambering around us. I tell them, "Hey boys, take your time. We're not in any hurry. We don't want anyone getting hurt out there." It is over.

After Gemini 10 there were two more flights in the series. Gemini 11, with Pete Conrad and Dick Gordon, launched in September 1966 and flew a very nice flight indeed. Their rendezvous employed a "brute force" technique, overtaking their Agena at high speed before the completion of their first orbit, as opposed to the leisurely four-orbit approach most other flights had used.

Just as John and I had effortlessly raised the world altitude record from Voskhod II's 167 miles to our 475, so did Gemini 11 break ours, soaring all the way up to 850 miles. Up there they took some spectacular photographs of the Persian Gulf, showing the earth's curvature in dramatic fashion. Conrad came through as a cool and competent commander without losing any of his cheerful, bubbly exuberance, and Dick Gordon's only problem came during EVA. Perched on the nose of the Gemini, he was trying to connect a tether between it and their docked Agena. It was a real wrestling match and all Conrad could do was yell, "Ride ' em, cowboy" to the sweating Gordon. Tired and overheated, Dick was ordered back to the cockpit shortly thereafter.

In view of Gordon's difficulties and Gene Cernan's, including very high heart rates, plus comments from Ed White and me, Chuck Mathews decided to use Gemini 12 to test some devices to make a spacewalker's tasks easier.

Jim Lovell was the commander and Buzz Aldrin the EVA guinea pig. Buzz was not too pleased because he had hoped to do a Buck Rogers zip-around-the-sky using a modified version of Cernan's AMU. Instead he had the less dramatic but equally useful task of evaluating a group of handholds and footrests, anchoring devices to hold the body in position while he practiced working with special tools. Buzz spent over 5 hours EVA in a cool and methodical demonstration, proving that an EVA was easy if proper body restraints were provided. Gemini 12 also included another successful rendezvous and a long list of experiments. It was an impressive flight with which to end a proud series.

Today the public's memory of Gemini is lost in Apollo's shadow. With six highly publicized Mercury flights preceding it, Gemini's accomplishments were not perceived as being dramatic improvements, yet those ten manned flights were absolutely essential in proving the technology of the machines and the durability of the humans.

Demonstrating various rendezvous concepts was a crucial precursor to a lunar voyage, but no less important was the 14-day epic journey of Borman and Lovell, an extraordinary marathon covering over 5 million miles in over 200 trips around the earth. Lesser accomplishments included extensive extravehicular operations, routine maneuvering in space, and guiding a spacecraft during entry to a precise landing point (the last seven Gemini flights landed within 7 miles of their target).

An overlooked but equally important benefit was the fact that the gap between Mercury and Apollo was filled in such a manner that program continuity was achieved. Ground personnel and flight crews gained experience and confidence in their ability as a team to solve complex, inflight problems; managers experienced budget and personnel stability; Congress was presented a logical step-by-step process for expanding to a lunar capability.

All in all, Project Gemini served as a bridge between the rudimentary Mercury capsule and the sophisticated Apollo spacecraft, a bridge between President Kennedy's bold statement and the national capability to execute it—a fine, sturdy, and necessary bridge. Today Chuck Mathews is one of the few who remembers that, and more. "Gemini was really a little space station," he muses. Certainly it was a bridge to much larger things to come.

4

Decade's End

The Saturn V moon rocket contained 3 million parts, the Apollo Command and Service Modules 2 million, and the Lunar Module 1 million. All 6 million worked, nearly all the time. In 1962 the Atlas was strained to its limit in putting Glenn's 3,000-pound Mercury capsule into space; less than 6 years later the Saturn V delivered 250,000 pounds to low earth orbit. The evolution of this gigantic pile of machinery, and how NASA made the quantum jump in power and performance, has its origins in the moon itself.

As the most prominent object in our night sky, the moon has always attracted the eye and the mind. Early Babylonian, Egyptian, and Chinese astronomers understood that the moon circles the earth once every 29½ days. In Western society the idea of an earth-orbiting moon was easy to accept because the popular notion, articulated by Ptolemy of Alexandria in the second century after Christ, was that *all* celestial objects revolved around the earth.

In the 16th century, the Polish astronomer and mathematician Nicholas Copernicus put the earth and other planets in their proper places, orbiting the sun, but kept the moon in a circular earth orbit. Johan Kepler refined the Copernican concept by describing the paths of moon and planets as ellipses rather than circles. Today we know that the lunar ellipse swings from an apogee of 253,000 miles to a perigee of 221,000. This orbit is "perturbed," the term astronomers use to describe deviations from a perfect path, by the gravitational pull of the sun and by the fact that the earth itself is not an exact sphere.

Having seen the moon, first with the naked eye and then through Galileo's telescope in the 17th century, people turned their attention to methods of getting there. Kepler himself speculated on the journey, writing a fanciful account of a sleeping man (himself?) transported to the moon by demons. By so doing he almost caused his mother to be executed as a witch.

My favorite lunaphile is Cyrano de Bergerac, who in 1649 wrote a novel, *The Voyage to the Moon.* His protagonist first tied vials of dew to his body, a propulsion system based on the lifting power of early-morning evaporation. That method failing, our hero finally reaches the moon by firecracker power, the first to use Sir Isaac Newton's third law (''for every action there is an equal and opposite reaction''), only 7 years after Sir Isaac was born. From space, the earth looked to de Bergerac like a large Holland cheese, which is better treatment than that given it by one of Mark Twain's characters, who referred to it contemptuously as ''the wart.'' But of all the fictional lunar voyages, the one that made the most lasting impression on our world was Jules Verne's fantasy, as described in his book *From Earth to the Moon.* His spacecraft, with a crew of three, was launched from Tampa, Florida, by a gigantic cannon named Columbiad, and returned after a lunar voyage to an ocean landing—amazingly close in some respects to an Apollo round-trip. Verne also discussed rocket power, escape velocity, and weightlessness.

By the early 20th century people were getting serious about reaching the moon. An obscure Russian schoolteacher named Konstantin Tsiolkovsky wrote seminal analyses of spaceflight, including the use of multistage rockets. Tsiolkovsky considered the moon just a way station on mankind's celestial migration, but apparently he felt a personal longing for it: ''to lift a stone from the moon with your hand. . . .'' In Germany it was Hermann Oberth, a Transylvanian mathematician who corresponded with Tsiolkovsky and who, like the Russian, was a brilliant theorist. Many of Oberth's ideas eventually proved correct, such as using hydrogen as a fuel (as Apollo did) and building lightweight structures that depended (like a balloon) on internal pressure to keep them rigid. The Atlas booster was the first practical application of this idea.

In America Robert Goddard combined theoretical genius with a tinkerer's yen for seeing the results of his dreaming. Goddard actually built liquid fuel rockets in Massachusetts, until he was hounded out of the state by an unfavorable press and unsympathetic safety officials. Later he tried again, near Roswell, New Mexico. Goddard wanted to send a payload of flash powder to the moon, where its impact would be visible through telescopes on earth.

In a 1920 editorial the *New York Times* sniffed at this suggestion, concluding '' . . . that will be believed when it is done.'' The basis for the *Times*'s skepticism was the editorial writer's utter failure to grasp the principle of rocket propulsion, believing that a rocket would only work in the atmosphere, where the exhaust gases would have something to push against. Sir Isaac Newton knew better: ''for every action, an equal and opposite reaction.'' In fact, rocket motors work more efficiently in the vacuum of space, with no air to impede the expansion of exhaust gases through the rocket's nozzle.

Although the Big Three—Tsiolkovsky, Oberth, and Goddard—understood full well the potential of the rocket, their governments did not—at least not until World War II. Prior to this, rockets had powered automobiles, gliders—even bicycles and roller skates—but it was the Germans at Peenemünde on the Baltic Coast who first built a long-range missile powered by a liquid fuel rocket. It was here that Wernher von Braun developed the V-2 team of rocket experts, most of whom emigrated to the United States in 1945.

The following year saw the RAND report and the burgeoning of interest in spaceflight—even to the moon. As early as 1959, the fledgling NASA was pondering the engineering problems associated with lunar flight but received scant support from the Eisenhower administration. All that changed in May 1961 with President Kennedy's bold invitation to the Congress: "I believe we should go to the moon." Tsiolkovsky died in 1935 and Goddard in 1945, but among those who watched Neil Armstrong, Buzz Aldrin, and Mike Collins depart for the moon on July 16, 1969, was Hermann Oberth!

Between Kennedy's decision and our execution of a moon landing, it was vital to learn more about our destination. Earth-bound telescopes could not resolve lunar objects smaller than a football stadium—not nearly enough detail to tell whether a landing could be safely made. What was the surface like? Was it flat enough for the Lunar Module, and strong enough to support its weight? Theories abounded, and several were not optimistic about our chances. One model of the moon's surface was of a fine dust layer, perhaps 50 feet thick, that would swallow a Lunar Module. A variation on this theme was a thinner coating of dust, but one with an electrostatic charge that would cause it to cling to a Lunar Module's windows, obscuring the view. Another suggestion was that the surface was composed of "fairy castle" material, a frozen, stony froth that would crumble upon impact. Finally, lava tubes were hypothesized to lie just beneath the surface, causing cavities that a landing gear could penetrate, overturning the Lunar Module.

To separate fact from fancy, NASA decided to send unmanned probes to the moon, to provide Apollo's designers with some hard data, not just about the properties of the surface, but also on gravitational effects, radiation hazards, and the presence of meteoroids. The first was Ranger, an 800-pound spacecraft designed to crash directly into the lunar surface, clicking a camera lens as it flew to destruction. The initial six tries were dismal failures, but the final three flights, in 1964 and 1965, returned thousands of photographs far superior to those produced by any telescope. The last couple of frames, taken a second or two before impact, were especially valuable in showing a surface pockmarked with craters of all sizes and littered with boulders large and small. It looked possible to land, but definitely not just *anywhere*.

The Russians were the first to soft land a probe on the moon, in February 1966, with the Luna 9. We followed closely with a series of Surveyors. These spacecraft, like Ranger, showed barren and battered plains, but of greatest importance was the fact that the surface easily supported their weight. The surface

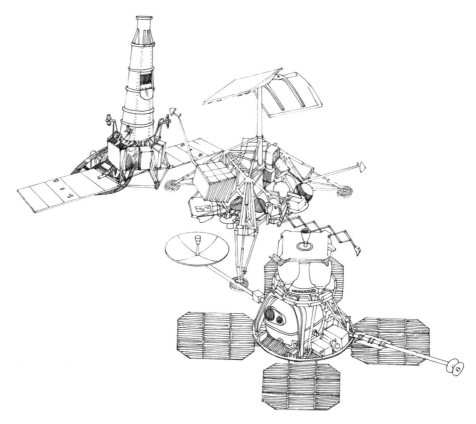

Ranger, Surveyor, and Lunar Orbiter

material seemed to be the consistency of damp sand and of a chemical composition resembling the family of volcanic rock called basalt.

The third and final type of pathfinder was Lunar Orbiter, which mapped the equatorial regions of the moon and provided planners with the data they needed to pinpoint ideal landing spots. As a bonus, Orbiter confirmed that radiation levels near the moon were low, and discovered that the gravitational pull of the moon was not uniform. This latter point was a mystery associated with buried mass concentrations, dubbed ''mascons,'' that caused slight dips (''perturbations'') in Orbiter's path as it passed over these spots.

Ranger, Surveyor, and Lunar Orbiter all provided clear evidence of intense meteoroid activity in the moon's past, but none detected a meteoroid in the act of hitting the moon. The coast seemed clear, but this was not known until 1967, after the Mercury and Gemini programs had ended, and by that time the Apollo designers had already committed their machines to production.

When President Kennedy gave the Apollo go-ahead, there were less than 9 years remaining in the decade, and NASA's only flight experience was Al Shepard's 15-minute suborbital lob. The only thing that was clear was that "landing a man on the moon, and returning him safely to earth" was going to require one hell of a big rocket, one which did not exist.

Fortunately, 2 years previously Wernher von Braun's group at Huntsville, Alabama, had been transferred from the Army to NASA. Since before Peenemünde days von Braun had been dreaming of spaceflight ("he aimed for the stars . . . and hit London," according to comedian Mort Sahl), and now was Wernher's chance to realize that dream.

Von Braun had been studying various booster options, mostly using clusters of small engines but also including a rocket powered by a single engine, a 1.5 million-pound thrust giant sponsored by the Air Force and known as the F-1. Von Braun's series was known as the Saturn program, a family name that included almost anything Huntsville could conjure up. Beyond Saturn there might be Nova—a cluster of as many as eight F-1s, although Nova gradually became subsumed by larger Saturns. For example, the Saturn IV had four F-1 engines, but was never built, while the Saturn V eventually became the approved moon rocket. Before it was built, however, smaller Saturn I and IB boosters were provided for earth orbital flights.

There were very good reasons for the myriads of Saturn and Nova designs. Von Braun knew where the moon was and what escape velocity was required to reach it, but what mass had to be accelerated to that velocity? The machines carrying the men had not been designed, nor could anyone even outline the mission profile. If a direct ascent and return by one spacecraft were employed, a 12 million-pound thrust Nova-class behemoth would be required. If the craft was assembled in earth orbit, two smaller boosters would be needed. If lunar orbit rendezvous was selected, perhaps one intermediate-sized rocket would suffice.

Before a trip to the moon, however, there would be preparatory flights in earth orbit, not to mention the fact that jumping from existing rockets to something as large as a Nova did not appear practical without an intermediate step. Before Huntsville became part of NASA, the Defense Department had ordered a clustered rocket, using eight existing engines, so that von Braun and his crew were working on what came to be called the Saturn I long before NASA had decided on lunar orbit rendezvous and the Saturn V design that emerged from analysis of such a mission.

One early decision made by NASA was the selection of hydrogen as the fuel for the upper stages of the Saturns. For years John Sloop and others at NASA's Lewis Research Center at Cleveland had been intrigued by the high specific impulse of liquid hydrogen as a rocket propellant. In theory it could operate 30-40% more efficiently than the Atlas fuel (kerosene) or the Titan II's (hydrazine). As early as 1903, Konstantin Tsiolkovsky had suggested liquid hydrogen as a fuel, as had Goddard and Oberth somewhat later, and the 1946 RAND report offered it as an alternative to kerosene. But it wasn't a simple decision.

As the designers of the ill-fated zeppelin *Hindenburg* could attest, hydrogen was a tricky, highly explosive gas. In liquid form it boils at minus 423°F, and keeping it below that temperature in storage tanks and feed lines is difficult. Also, liquid hydrogen is very light, only 7% as dense as water, and requires huge tanks compared to conventional fuels. Especially in the first stage of a rocket, tank diameter may grow unacceptably large. Weighing all the pros and cons, and sticking their necks way out, a committee headed by Lewis's director, Abe Silverstein, opted for liquid hydrogen/liquid oxygen engines for all the

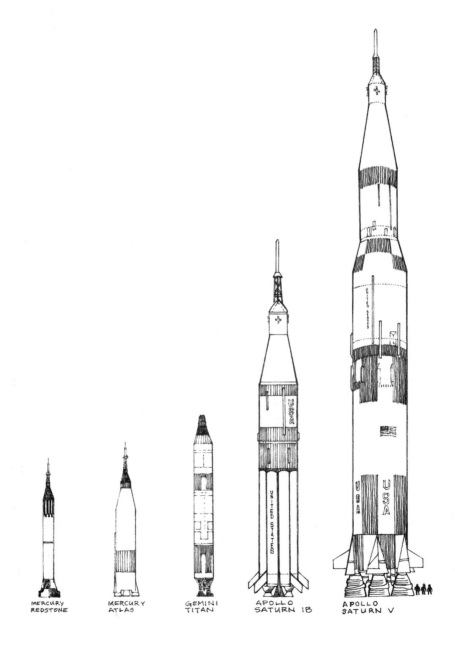

MERCURY   MERCURY   GEMINI   APOLLO   APOLLO
REDSTONE   ATLAS   TITAN   SATURN 1B   SATURN V

Saturn's upper stages. At first a skeptic, von Braun soon became an enthusiastic hydrogen supporter. In retrospect, the Silverstein committee's decision was not only brave but wise, and resulted in the greatest advance in rocket technology since von Braun's World War II V-2.

The choice of the Lunar Orbit Rendezvous mode was, like many NASA decisions, one that boiled up through NASA's ranks. John Houbolt at Langley was the first champion of LOR, and he convinced Bob Gilruth. Gilruth's people, in turn, sold it to the von Braun team at Huntsville, and both centers then ganged up on Washington headquarters, which had been holding out for direct ascent via a gigantic Nova. Jim Webb became converted to lunar rendezvous, and gave his approval, despite stubborn opposition from the President's science advisors. The argument for direct ascent was that it seemed the most simple—just one machine, there and back. But the size of that machine! Not just at launch, but at the moon as well.

Consider streaking toward the lunar surface in something larger than a locomotive, everyone on board bug-eyed at the blur of craters hurtling toward them, committed by their trajectory to some kind of a landing, yet still unsure whether the braking rockets will work to perfection and ease them to a stop at precisely the last, and right, instant. LOR, on the other hand, allowed the use of a tailored spacecraft, a small, maneuverable landing module that could leave the locomotive up above and then return to it.

With a firm commitment to LOR, NASA planners could proceed logically with the design of the various machines. Huntsville was unsure whether a four- or five-engined Saturn would be required, and wisely picked the larger design, allowing some margin for the inevitable growth of payload weight. A mother ship, or Command Module, would make the round-trip. Attached to it would be the storehouse, the Service Module. Together these two were known as the Command and Service Module, or CSM. The Lunar Module was a two-stage vehicle, the bottom half, or descent stage, acting as a lunar launch pad for the ascent stage, which returned to the CSM. The CSM held a crew of three, two of whom landed in the Lunar Module. The Command Module would be on top of the stack, with an escape rocket attached to it in case of trouble on the launch pad or early in flight. The Saturn V would have three stages, the upper second and third stages using hydrogen but the first stage sticking with kerosene because of its smaller tank size.

That was the outline of the plan in 1962, and it remained unchanged, a sturdy skeleton upon which NASA built for the next half dozen years. Simultaneously, Project Gemini was being developed to probe problems of rendezvous and long duration spaceflight, but although Gemini caused innumerable revisions and refinements of technique, and answered many questions, the basic Apollo machines were well conceived and proceeded in parallel with Gemini.

My introduction to Project Apollo came at Edwards Air Force Base, although it took my wife and me a while to realize it. Across the dry lake bed, some

15 miles from where we lived, there was a hillside lined with engine test stands. Known simply as "the rocket site," this installation produced gargantuan belches, crackling rattles, and great whooshing roars—usually in the middle of the night. One sound in particular exceeded all others in intensity, rolling across the lake bed like a tidal wave, rumbling right through our bedroom windows and tinkling every teacup in our kitchen.

Upon inquiry, I found that the teacup tinkler was the F-1, 1.5 million pounds of thrust energy. The Rocketdyne division of North American Aviation, the F-1's builder under an Air Force contract, had an engine test facility in the Santa Susana mountains north of Los Angeles, but the king-sized F-1 had been moved to Edwards in the Mojave desert because of safety and noise considerations. Apparently we Edwardsites didn't count, and it *was* true we were accustomed to loud aircraft noises. Sonic booms by day, rocket blasts at night, Pat and I were glad when we moved to Houston, where there were more people but less noise.

Von Braun had decided that the Saturn V was too big a project to trust to one manufacturer. Dividing up the construction work made good sense from a technical point of view and was even better politically, as the entire aerospace industry was understandably eager to participate somehow in this technological and financial bonanza.

The Boeing Company was the successful bidder on the first stage, 33 feet in diameter and containing five F-1 engines fed by gigantic kerosene and liquid oxygen tanks. North American won the second stage, with five hydrogen-burning J-2 engines, and Douglas Aircraft the third stage, powered by a single J-2. The inertial guidance system and computer were housed in a ring mounted atop the third stage and called the Instrument Unit, developed by von Braun's own people at Huntsville and later turned over to IBM. Also at Huntsville was the test stand that checked out the first stage with all five engines churning away. Together the F-1s produced 7.5 million pounds of thrust, or about 180 million horsepower. Of this, approximately 2 million horsepower was converted into noise.

The noise that Jim Webb heard in Washington was of a different sort but just as loud. The total cost of the Apollo program was estimated to be between $20 and $40 billion. Congress had greeted President Kennedy's proposal with quick approval, and now waited for contracts to be awarded. Jim Webb knew that he needed an iron-clad system to defend his major procurement decisions against criticism from Congress, the press, and disgruntled contractors. Webb insisted on a personally approved, comprehensive procurement plan before issuing a request for bids. Included in the plan was a point-scoring system in which relative weights were assigned to such factors as cost, prior experience, management skill, and technical merit. Two separate groups assessed technical and business factors, and reported to an evaluation board, which, in turn, gave a comprehensive briefing to Webb and his top two assistants.

The noncompetitive award to Douglas of the third stage of the Saturn V, called

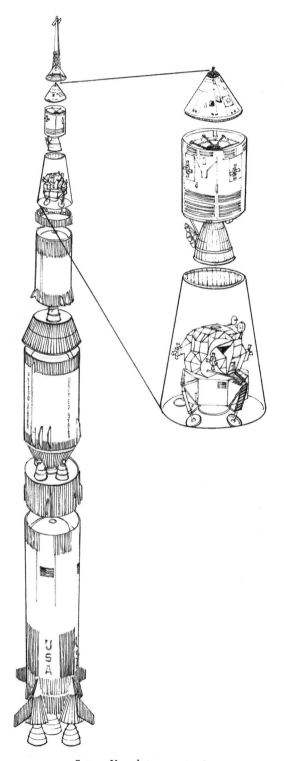

Saturn V rocket components

the S-IVB, was an exception to this procurement process. Douglas already had a contract to produce the S-IV, the upper stage of the Saturn I, the earth-orbiting predecessor to the Saturn V. (The nomenclature of these stages is confusing because of the early uncertainty of how many rockets of various sizes might be required to reach the moon using different rendezvous techniques. At one point four stages were thought necessary, therefore the fourth was to be called the S-IV, followed by a more powerful version, the S-IVB. These names stuck, even after the fourth stage was found unnecessary, using LOR as the mission-planning bible.) The S-IVB, like its predecessor the S-IV and the earlier Centaur, used liquid hydrogen, but unlike them it would be "man-rated," as NASA put it. Because of Douglas' earlier experience, the S-IVB was also the first of the three moon rocket stages to emerge and it received a great deal of early management attention from Huntsville.

Traditionally there are four stages of development—hence problems—with a launch vehicle. First is the basic design, which always contains some unknowns. Second is the engine, which may work fine on the test stand (especially in the middle of the night at Edwards), but must be mated with the rest of the booster and be made compatible with tanks, feed lines, etc. Third is the production phase, which inevitably reveals flaws either in the design or in execution of the design. And fourth is flying the vehicle—it must be tested in its actual environment. There were plenty of problems in the first three development stages of Saturn V, but the fourth stage was remarkably trouble free. There were 32 Saturn launches in all, without a single mission failure. Thoroughness in ground testing paid off in flight.

The rocket motors themselves, the F-1s of the first stage, and the J-2s of the second and third stages, both encountered early problems. With the J-2 the problem was the injector, the device for introducing a fine mixture of fuel and oxidizer into the thrust chamber. The combustion of hydrogen and oxygen was producing so much heat that the engine was literally consuming itself, burning up the injector within seconds. A cooler, porous-faced injector was substituted. With the F-1 the problem was more traditional: combustion instability. Just as an automobile engine can knock, so can a rocket motor experience unsteady combustion. In a car engine knocking is not catastrophic; in a rocket, however, combustion instability can cause the thrust chamber to disintegrate.

One critical factor was the rocket engine's ability to recover from a momentary abnormality and restore itself to safe, stable combustion. But how to determine this on the test stand? Clever engineers induced combustion instability by using "bombs," explosive charges deliberately set off inside a thrust chamber to see how the engine would react to a near-instantaneous interruption of smooth, controlled burning. Each bomb was wrapped in a protective cover so that the engine had a chance to start and reach full power before the nylon cover melted and the bomb ignited from engine heat. These bombs were tiny, as small as 50 grains, but their explosions were still effective in disturbing the combustion process and creating a realistic indication of the chamber's stability.

In 1962 three F-1 engines were lost to combustion instability before a redesign of the injector solved the problem. The injector was built of copper rings with holes through which fuel and oxidizer squirted. In something as big as the F-1, the injector was likewise of mammoth proportions; where an automobile might have one injector orifice per cylinder, the F-1 injector had 3,700 orifices for fuel and 2,600 for oxidizer. Tinkering with this device, the very soul of the machine, or anyway the part least well understood, was not a precise operation. Trial-and-error methods were required, and the final solution, which included enlarging the fuel orifices and changing the angle at which fuel and oxidizer jets converged, was the result of 18 months of intensive experimentation. In the end, a bomb could be exploded inside an F-1 engine and it would recover within one-tenth of a second.

Another major hurdle with the F-1 was pumping sufficient fuel and oxidizer to the thrust chamber. The turbopump had to deliver 15,000 gallons of fuel and 25,000 gallons of oxidizer per minute! A 55,000-horsepower turbine was needed to drive the pump. Engineers did all they could to keep these machines as simple as possible, and to rely on previous designs, but the size of this equipment alone introduced a host of new problems. About a dozen turbopumps failed in ground tests, for a variety of reasons, but the Rocketdyne and NASA engineers doggedly kept at it until gradually their problems disappeared. For example, they had chosen a new nickel steel alloy, Rene 41, for part of the turbine, for the same reasons the Mercury engineers chose it to protect their spacecraft against entry heating. They found it was a wonderful high temperature material but they couldn't weld it without causing cracking. Welders had to go back to school before the F-1 engine could fly.

Once the J-2 and F-1 engines had demonstrated their reliability in ground tests, they had to be integrated into the overall launch vehicle design. Liquid hydrogen was proving to be a significant problem there, as expected. Douglas Aircraft was unsure of what material to use to insulate the third stage hydrogen tank. Originally they thought balsa wood was superior to any man-made material, but an analysis of the available supply showed that every balsa tree in South America would have to be denuded to supply sufficient insulation for the Saturn fleet. Instead, polyurethane was poured into fiberglass forms to produce waffle-like tiles that were glued in place *inside* each hydrogen tank—43,000 tiles per tank. On the second stage North American engineers placed their insulation on the *outside* of the tank. Inherently this design seemed simpler, but in fact the extremely cold surface of the hydrogen tank caused chunks of insulation to come unglued and fall off.

Once a tank was filled with hydrogen it had to be inspected for leaks, and that was a tricky procedure. A jet of escaping hydrogen was apt to combine with oxygen in the air and burn, but in daylight the flame would be invisible. Douglas used infrared cameras on its test stands to detect hydrogen leaks, but the cameras had blind spots, so men were sent around the scaffolding, wearing protective clothing and holding brooms out in front of them. If a broom suddenly burst into flame, there was a leak! Two stages, one S-IV and one S-IVB, did blow up on the Douglas test stand, but oddly enough hydrogen was not a culprit in either case.

If Douglas was having problems, they were mild compared to North American's on the S-II. Not only was insulation falling off the second stage's hydrogen tank, but welding defects seemed to make manufacture of the tank nearly impossible. For building the S-II North American had picked an aluminum alloy that got stronger as it cooled to cryogenic temperatures, but the material was difficult to weld. The tank circumference was slightly over 100 feet and a flawless curved weld of this length (where the tank dome joined the cylindrical main body) seemed beyond North American's grasp.

"The S-II is out of control. . . . It is apparent that . . . management has not been effective. . . . There are still significant technical difficulties . . .'' reported one of von Braun's managers. NASA headquarters sent Air Force General Sam Phillips out to California to investigate the S-II stage and North American's other responsibility, the Apollo Command and Service Module. His gloomy report ripped into the company—the first lunar landing might not be possible in the decade unless North American mended not only its welds but its corporate ways. North American quickly responded with various personnel changes and reorganizations, and NASA continued to shower the S-II program with attention and advice. From 1965 to 1967 great strides forward were taken and the S-II, like the stage above and below it, compiled a perfect record in flight.

At the base of the Saturn V stack was the first stage, Boeing's S-IC. By far the largest, it was also the easiest to develop (not counting the difficulties encountered by its five F-1 engines). This can be traced partly to a very close early working relationship between von Braun's engineers and Boeing's, and partly to the selection of kerosene and liquid oxygen propellants, whose properties were well known. Still, with a tank diameter of 33 feet, and with engines gulping propellant at over 3,000 gallons per *second*, it was surprising that Boeing did not unearth some gargantuan problems along the way.

Von Braun's philosophy was a mixture of the Army's arsenal system and the Air Force's contracting methods. Like the Army, he believed in dirty-hands engineering, and he wanted his people to understand every step in the design and manufacturing process. However, he recognized that with a project as complex as the Saturn it was necessary to rely heavily on industry, as the Air Force did. Looking back on earlier and simpler days, von Braun noted in 1962 "our rocket team has become today more than ever a managerial group."

This managerial group sometimes seemed to envy its contractors, the people who were actually building these wonderful machines. Certainly Huntsville's control over contractors was tight, tighter than most of them were accustomed to or appreciated. Von Braun himself, however, was far from a martinet. He was a friendly, charming man, with a rare flair for explaining complex machines in picturesque language a reporter or a Congressman would remember. He kept up with the technical work of his subordinates and frequently contributed to detailed technical decisions. He was not seen as a meddler, however, but as a leader who knew his work and his workers, and who could transmit to them his own cheery enthusiasm.

Von Braun's people felt they were part of a skilled, cohesive team and their loyalty to their boss far exceeded that of most civil servants. Von Braun's personal charisma was strong, but it alone was insufficient to cause the 3 million Saturn V parts to assemble themselves and find their way to Cape Kennedy in time for launch. At its peak the Saturn program was a model of organization, a multifaceted operation that scheduled barges through the Panama Canal with the same precision that it X-rayed titanium welds.

At no point was the precision of NASA's team more apparent than at the Cape on a launch day. A Saturn V on the launch pad is a solitary sight, but in fact it is linked to its builders by several thousand data points. Every critical component's temperature or pressure is monitored constantly, and the results displayed in summary form on more than a hundred computer terminals in the firing room at the Cape. If all the data coming from a Saturn V were printed, the information would fill 300 pages each second.

The first Saturn V, called 501, was launched November 9, 1967. Originally, von Braun had planned it as a test of the first stage only, with inert second and third stages. But 501 contained not only three live stages but an actual CSM as well. This change could be traced to Dr. George Mueller, a key aide to Jim Webb. Several years earlier, Mueller had startled the daylights out of the Huntsville group by introducing what he called ''all-up'' testing. The conservative approach of testing one component at a time was too expensive and time-consuming, felt Mueller, and would never get us to the moon by the end of the decade. Careful ground tests of the parts should be conducted, but when it came time to fly—go for it, ''all-up,'' in a form as close to the final configuration as possible. Mueller had converted von Braun to this new concept, and now both waited for 501 to prove them savants or fools.

I watched 501 from a causeway 4 miles from Launch Pad 39, as close as the safety people would allow. At that distance the monster looked like a thick white pencil, giving off dainty wisps of steam in the thin November sunshine. As the engines ignited, telemetry data streamed into the firing room. When the five F-1s reached full thrust, and the experts were assured that all was well, the rocket was released and began a slow, stately climb.

At ignition the flame was red-orange, but rapidly changed to an incandescent white at its core and a dirty brown at its edges. The scene had an eerie quality because for the first 20 seconds it occurred in total silence. When the sound wave reached us with a sudden jolt, it was more than just a noise. The sand under my feet began vibrating and I felt as if a giant had grabbed my shirtfront and started shaking. Gradually the noise subsided as the rocket, picking up speed, arced off to the east. I found out later that all three stages had functioned perfectly, giving George Mueller solid proof that, in this case at least, all-up testing had matched his most optimistic estimates. Almost as a bonus, the Saturn had propelled the CSM to an altitude of 9,700 miles and a speed of 25,000 mph, thus generating an amount of heat roughly equivalent to that of a lunar return trajectory. Although the CSM's heatshield reached 5200°F, temperatures inside the cockpit were easily within the limits of human comfort.

502 was a different story. Something went wrong with all three stages. First the S-IC developed a Pogo, the same longitudinal oscillation that had plagued the Gemini program. Then two of the five J-2 engines on the S-II stage shut down and the third stage barely limped up into a lopsided orbit. Finally the single J-2 on the S-IVB refused to reignite to simulate departure from the moon. Shocked by this debacle in the aftermath of 501's success, von Braun appointed investigation teams, and one by one the failures were analyzed and fixed.

The oscillation in the first stage was cured by using pressurized helium gas as a shock absorber in the liquid oxygen lines. There were two separate problems with the second stage. First a fuel line had ruptured, spraying out hydrogen. A detection circuit sensed this and sent a shutdown signal—to the wrong engine. Thus two engines were lost. Fixing the electrical problem was easy but it took a bit longer to discover that a high frequency vibration, or "buzzing," in the fuel lines was causing metal fatigue and subsequent failure. The buzzing was eliminated by adding bends in the lines, a fix that solved the J-2 problem in the S-IVB stage as well as the S-II.

Then on to 503, but what kind of a flight should it be? To understand how that was decided, it is necessary to explore the development of CSMs.

Back in November 1961, the CSM contract was awarded to North American Aviation (later known as North American Rockwell, and today simply as Rockwell International). At that time, only 6 months after Kennedy's call for a visit to the moon, it was not clear exactly what equipment was required to land on the moon. The name Apollo, suggested by Abe Silverstein, had been approved, but it was a vague term, an umbrella covering any planned activities after Project Mercury. The CSM was to go to the moon, but at that time its precise role was hazy. For instance, the CSM might hook up with either a "lunar landing module" or a "space laboratory module." It might land on the moon itself, attached to a landing module, or it might orbit overhead while a separate craft landed.

In any event the Command Module would be required to return to earth, enter the atmosphere at 25,000 mph, and then land with a crew of three. That much was known, and that was plenty to get North American started on design work.

Bob Gilruth's team, especially Max Faget and Caldwell Johnson, had strong ideas about the CSM. Based on their Mercury experience, and their earlier work in the Pilotless Aircraft Division at Langley, they believed spacecraft should be symmetrical about the long axis of the booster, or "axisymmetric" as they called it. Unsymmetrical shapes were much more difficult to control at high speeds. Also, a spacecraft had to be compact, not a fancy shape difficult to manufacture or to protect from entry heating. Further, it had to accommodate a three-man crew, for the simple reason that a 24-hour day in space could best be divided into three 8-hour shifts. So a three-man, compact, axisymmetric shape was the starting point.

Next consider certain other basic facts. The gravitational pull of the earth and the distance between earth and moon would result in a returning spacecraft hitting the atmosphere at 25,000 mph. This was easily calculated, but not so well known was how a spacecraft could be slowed from this unprecedented speed without skipping back out of the atmosphere. Or, once captured by earth's gravity, how the spacecraft could be guided to a safe landing at a predetermined spot. The earth was rotating 360° in 24 hours, so the landing point itself was moving at considerable velocity; if a returning spacecraft was 5 minutes early or late it would land over 50 miles east or west of its target, unless some adjustment to its flight path could be made.

Unless some adjustment could be made! That was the key, and it had to be a large adjustment, not just to compensate for timing errors but also to respond to changes in plan, such as moving the landing point to avoid a raging hurricane. The size of the landing "footprint," as it was called, depended on how maneuverable the spacecraft was. In other words, its lift to drag ratio. Faget had intentionally designed the Mercury capsule with an L/D = 0, figuring that an inflexible flight path between retrofire and splashdown was desirable. But to play the lunar game a different set of rules was required, and some lift was definitely needed. As Faget put it: "During the initial phases of entry, the spacecraft must pull negative lift to skim along a very narrow corridor of the upper layer of the earth's atmosphere. If the upper layer of this corridor were exceeded, the spacecraft would skip out of the atmosphere back into a highly eccentric orbit and perhaps expend its supplies before entering the atmosphere a second time. On the other hand, if the corridor boundary were missed on the lower side, the spacecraft would exceed the heating or load limitations of its structure."

The precise value of an adequate lift to drag ratio was hotly debated within NASA and industry. Much depended on how accurately the guidance and navigation system could steer the spacecraft to the center point of the entry corridor. Normally this task would fall to gigantic earth-bound radars and computers, but mission planners worried about a communications failure, in which case the crew would have to do their own navigation. If they erred and hit the entry corridor high or low, some extra lift would be invaluable in compensating for their error, preventing either overheating or skipping out, and allowing them to reach a safe landing spot.

Eventually the experts picked an L/D = 0.5. Compared to airplanes, this is a tiny number. For example, modern sailplanes can achieve a lift to drag ratio 100 times as large. But on the other hand, for a spacecraft 0.5 is not bad, considerably better than Gemini's 0.15. From Faget's point of view, 0.5 was good, small enough that the Apollo Command Module could be another high drag blunt body, a derivative of the Mercury shape. The diameter of the CSM was determined not only by the requirement to carry three men, but also by the expected diameter of the upper stage of the moon rocket. To achieve an L/D = 0.5, the trim angle, or angle of attack, would have to be 33°. To get the spacecraft to fly at this angle, the center of gravity would have to be offset 7½ inches from the centerline, or "idiot point" as Faget's engineers called it.

These details fell into place once the all-important L/D was known. The Command Module shape was not identical to that of the Mercury capsule, because at 33° the Mercury afterbody would protrude too far out into the slipstream and overheat. Therefore the shape of the Apollo Command Module had a more highly tapered afterbody—more like a gumdrop, the name bestowed by the Apollo 9 crew on their spacecraft.

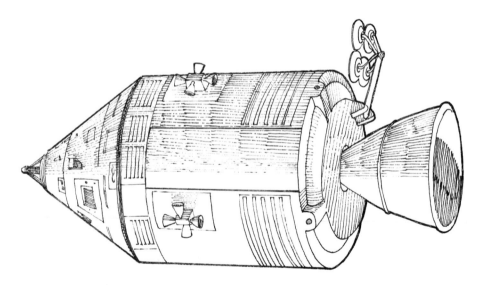

Command and Service Module with Service Propulsion System

In addition to specifying to North American the basic size and shape of Gumdrop, NASA also spelled out the equipment it had to contain. Like the adapter section of the Gemini, the Service Module was the storehouse and was dropped off before entry. It was to have three fuel cells, producing electricity and drinking water by combining hydrogen and oxygen. It also contained the attitude control system and a single, large rocket engine, the service propulsion system (SPS).

Since no parts of the CSM were superfluous, it is difficult to single out any one critical component, but somehow the SPS always seemed near and dear to my heart. This stubby cone sticking out the back end of the Service Module was what slowed the spacecraft as it approached the moon, but, more to the point, it was what broke the bond of lunar gravity and propelled the CSM back into an earth-return trajectory. It had redundant pumps and valves, but still . . . there was only one motor, one thrust chamber, one exhaust nozzle, and if it exploded or fell off or something, there you were for the rest of your life in lunar orbit. It was the equivalent of Gemini's four solid retrorockets but all together in a single package. The SPS was probably my all-time favorite piece of machinery.

The Command Module was a cone 13 feet in diameter and 11 feet high. The interior volume was four times that of Gemini, and those of us who had been shoehorned into a Gemini for a couple of days really appreciated the extra room. In other ways as well, the Command Module was clearly a more advanced machine, especially in the guidance and navigation department. Instead of the hand-held sextant and crude computer I had used on Gemini, the Command Module had a sophisticated navigator's station equipped with a sextant and a telescope and a much larger computer that had stored in it the coordinates of 37 stars. Just align the cross hairs of the sextant on one of these stars, push a button, tell the computer the number of the star, and it took over from there. It made me wonder why, with this design in the wings, we had been fooling around with Gemini's crude navigational scheme.

The computer could also be used to assist in maneuvering the spacecraft. Simply enter the direction in which you wanted to point, by giving it roll, pitch, and yaw angles, and it figured out how to use the least amount of fuel getting there, automatically firing the appropriate reaction control system thrusters. The guidance and navigation system was developed under a separate NASA contract by the Instrumentation Laboratory of the Massachusetts Institute of Technology. Its founder, C. Stark Draper, was a pioneer in inertial navigation, and today the laboratory bears his name.

The inertial measurement unit was a sphere slightly larger than a basketball containing three gyroscopes mounted at right angles to each other and connected to the spacecraft by swiveling devices called gimbals. The gyroscopes spinning at full speed kept the sphere pointed in one direction, or fixed in inertial space. Think of it as a "stable table" around which the spacecraft moved. Or think of the Greek words: *gyros* (turn) and *skopein* (to view), thus gyroscope—to view the turning. By comparing the position of the stars with the position of the stable table, and noting the gimbal angles, it was possible to calculate the direction in which the spacecraft was pointed. This procedure did not explain *where* in space the craft was located, only the *direction* it was pointed.

The *where* was provided by three accelerometers that sensed motion, and by a mathematical abstraction stored in the computer and called the state vector. A state vector is composed of seven numbers: three for position, three for velocity, and one for time. These are expressed as x, y, z; $\dot{x}$, $\dot{y}$, $\dot{z}$, and t. In other words, the state vector says that at some instant in time (t), the spacecraft was—measured from a reference point—at a distance x, y, and z from that point and traveling at a speed of $\dot{x}$, $\dot{y}$, and $\dot{z}$, x, y, and z are simply notations for distance divided into three components ("go north on Fifth Avenue seven blocks, turn east on 47th Street for two blocks, go up to the 35th floor"). $\dot{x}$, $\dot{y}$, and $\dot{z}$, the time derivatives of x, y, and z, are velocities expressed in the same coordinate system.

With an inertial measurement unit, a computer with a state vector, and a sextant to update this information, an astronaut had all he needed to navigate to the moon. In actual practice, the state vector was loaded into the Command Module computer on the launch pad and updated periodically inflight by information telemetered from the ground. In the event of the loss of communications, however, the astronaut could provide a less accurate but still acceptable update of the state vector by using the sextant. Updating the inertial measurement unit could only be done on board, shooting two stars with the sextant or telescope.

Atop the Command Module was something familiar to Mercury veterans: a solid rocket escape tower. The NASA designers had never been too comfortable with Gemini's ejection seats and, in any case, ejection seats for three astronauts would be heavy and difficult. Besides, any rocket big enough to fly to the moon could generate a fireball too large to allow escape with even the most powerful ejection seats. Therefore it was back to the old reliable, the solid rocket motor, although Caldwell Johnson grumbled years later that "it soaked up more than its share of engineering talent and imposed severe design requirements of its own on other parts of the spacecraft."

Also familiar were parachutes for landing, although this time there were three of them, each 88 feet in diameter. Originally, places like Australia and the western United States were considered prime landing sites but NASA soon switched to water. Water was flat, soft, abundant in the right places (close to the equator), and the Navy seemed eager to participate in the program and pick up returning moon men. Still, in the unlikely event of a landing on terra firma, the astronauts were supposed to be able to walk away from it (although some of us doubted this). The key to it was that the three couches were attached to the inside of the Command Module by telescoping struts. Inside each strut was an aluminum honeycomb material. The shock of a land impact was to be attenuated by crushing the honeycomb. Ingenious and proven in many ground tests but still . . . somehow the idea didn't appeal to us astronauts.

The fuel cells were also of some concern, because on Gemini they had been temperamental, and the water they produced had been undrinkable. Potable water from the CSM cells was a must if the stringent weight budget was to be met. Throughout the Apollo program, as in Mercury and Gemini, minimizing weight was a constant fight. Like compulsive eaters, the machines seem to grow inexorably and there were always plenty of excuses to explain why.

For example, sometimes testing produced unexpected results. When dropped into water the Command Module might float properly, right side up, or it might flip over and stubbornly remain inverted, submerging the crew's egress hatch. Obviously this situation had to be fixed and the solution involved adding air bags to the apex of the cone. When they felt their craft turning over, the crew would inflate the bags, causing the Command Module to pop up to an upright position.

Weight was usually added in the name of enhanced reliability. Everyone— from engineers to astronauts to project managers—was aware that the moon was a long, long way off. An emergency that would be tolerable in earth orbit, an oxygen leak, for example, could be fatal if it occurred 3 days from home. The Apollo equipment simply had to last or, if something broke, it had to be backed up by a similar unit.

Reliability also led North American and NASA down a primrose path called inflight maintenance. If an item of equipment broke, let the crew fix it. After all, the Mercury program had seen the metamorphosis of the astronaut from talking monkey to a contributing component of the design, and on Gemini the machine had been designed to be totally dependent on a human presence. Why not take it one step further and make the astronaut a repairman as well? This logic dictated that the equipment be located where the crew could get their hands on it, i.e., inside the Command Module. At the same time Mercury experience with a layer-cake stacking of equipment inside the capsule had not been satisfactory and Gemini had been designed to put as much machinery as possible outside the pressure shell.

CSM inversion

Apollo's designers were caught in the middle, and compromised. Most of the big stuff, as on Gemini, was contained externally, back in the Service Module, but lots of smaller items were stuffed inside the Command Module. As time went by it became apparent that most of these could not be repaired except at the factory, not unless the crew was provided with a large stock of spare parts, tools, and special repair facilities. The crew was handy in diagnosing trouble but could rarely fix it. Hermetically sealed units were more reliable and often two of them could be provided at a lighter weight than spares and tools.

Another mistake was the understandable eagerness of both North American and NASA to get on with it, not to waste a minute, to proceed with design and manufacture of the CSM. Even though details of the lunar landing mission had not been worked out, assumptions were made, blueprints drawn, and metal cut. This resulted in a Command Module design that had no docking mechanism or tunnel for entering a lunar lander, because it was designed at a time when the Command Module itself was a candidate for landing on the moon. This haste to rush into manufacturing caused wasted effort and money, equipment installed and then removed, wiring changes galore, and general instability and confusion about the Command Module's ultimate capabilities.

The first few Command Modules came to be known as Block I spacecraft, to be flown in earth orbit only, and to be followed by the real McCoy, the Block II lunar version. During development both modules grew so much in weight—nearly 50%—that it was no longer possible to keep the center of gravity offset enough to produce an $L/D = 0.5$. There was so much heavy equipment on board that it could not all be piled on one side of the spacecraft. However, as the L/D came down, it was offset by increased confidence in the navigation equipment. The more precise the entry into the atmosphere, the less lift was required. The L/D finally stabilized at 0.35.

Despite the growth in weight and complexity, the Command Module still retained its external dimensions, the compact, axisymmetric shape that Faget and his people had envisioned. Gumdrop was a dense package, with over 300 switches of one type alone, not to mention scores of levers, knobs, dials, valves, and other controls. It also contained hundreds of loose items: survival kit, food, sleeping bags, lithium hydroxide canisters to remove carbon dioxide, chlorine ampules to purify the drinking water, cameras, a medical kit, three pressure suits, star charts, checklists, emergency procedures books, extra underwear, spoons, toothbrushes, urine bags, and towels—and empty spaces to be filled with boxes of rocks from the moon. Gumdrop retained the Mercury and Gemini choice of atmosphere: pure oxygen at 5 psi.

North American assembled the CSM at Downey, California, near Los Angeles. Here parts were shipped in from all over the country, from hundreds of subcontractors. Here astronauts gathered to help in the design of the cockpit and the testing of critical components. It was similar to the McDonnell plant in St. Louis, except on a vaster scale, with huge clean rooms and more elaborate facilities in general. With Rocketdyne manufacturing the Saturn engines nearby, and Douglas working on the S-IVB in Long Beach, not to mention a separate North American plant for the Saturn S-II stage, southern California was the hub of Apollo activity.

Newly arrived from the Gemini program, I was impressed by the sophistication of the Apollo machines, but somehow disquieted by the attitude of those putting them together. The managers, who were supposed to have absorbed the lessons of Gemini, instead either ignored or disparaged that smaller program. The workers, unlike their St. Louis counterparts, seemed distracted by the pleasures of mountains and seashore. Years later I asked Joe Shea, who was the Apollo spacecraft program manager at the time, how he judged conditions at Downey.

"Well, quality versus money was always a problem. When North American won their contract, their managers threw a party and issued blue hats with NASA printed on them, only the S in NASA was a dollar sign: NA$A. I've still got one of those hats."

Nonetheless, despite the confusions of Block I design and escalating costs, both North American and Joe Shea seemed to be on top of things, almost cocky.

All that changed abruptly the evening of January 27, 1967, at Cape Canaveral. The crew of the first manned Block I Command Module, Gus Grissom, Ed White, and Roger Chaffee, were conducting a test on the launch pad when fire broke out inside their spacecraft. Within seconds the interior filled with flames and smoke. The temperature soared to 2,500°F. So rapidly did the combustion spread that the spacecraft overpressurized and burst a seam. The crew had no chance to escape but died of smoke inhalation and burns within moments. Joe Shea almost burned up with them. For some time he had been listening to Grissom complaining that the tests had not been proceeding smoothly. "Come on, sit in the lower equipment bay, and see how bad it is," Gus invited. Joe had accepted, but the North American people were not able to jury-rig communications for him swiftly enough, so he told Gus at breakfast that morning that he had opted for a later test: "If you think I'm going to sit at your feet for 4 hours and not be able to talk back, you're crazy."

The fire began in the lower equipment bay where Joe Shea would have been sitting. The exact cause was never discovered, but most likely was a short circuit that ignited combustible material that burned fiercely in the 100% oxygen atmosphere.

In space, pure oxygen at 5 psi is a manageable hazard. But on the launch pad NASA had routinely filled the Mercury, Gemini, and now the Apollo spacecraft with oxygen at slightly higher than ambient pressure. At 16 psi, oxygen greatly accelerates the combustion process—an entire cigarette, for example, burns to ash in a couple of seconds. And there was plenty of combustible material inside Gus's spacecraft: books, clothing, plastic netting, velcro patches, rubber insulation. Even steel can be set on fire at 16 psi, and some nonmetallic objects practically explode at that pressure.

The cause of the short circuit was never pinpointed, but it became abundantly clear that there were plenty of possible sources of a stray spark. The paperwork documenting the frequent changes to the Block I spacecraft had been sloppily maintained, and the accident board was never able to satisfy itself that it knew precisely which wires were inside at the time of the fire, which had been removed, which had been capped, etc.

George Low, who replaced Joe Shea as Apollo Spacecraft Program Manager after the fire, has written about the importance of this paperwork: "When machinery gets as complex as the Apollo spacecraft, no single person can keep all of its details in his head. Paper, therefore, becomes of paramount importance: paper to record the exact configuration; paper to list every nut and bolt and tube and wire; paper to record the precise size, shape, constitution, history, and pedigree of every piece and every part. The paper tells where it was made, who made it, which batch of raw materials was used, how it was tested, and how it performed. Paper becomes particularly important when a change is made, and changes must be made whenever design, engineering, and development proceed simultaneously as they did in Apollo."

The final nail in the crew's coffin was the design of the hatch. Even with a fireball at their feet, Grissom, White, and Chaffee might have escaped scorched but alive if they could have quickly popped the hatch open. But the Block I hatch was a primitive design, two separate hatches actually, and the inner, pressure-sealing one required the removal of several dozen bolts with a wrench before it could be pulled inside the spacecraft, exposing the outer, thermal-protective hatch, which was easier to remove. The inner hatch opened inward because the design was light and simple, and the pressure inside the spacecraft helped seal it. A hinged, outward opening hatch would have been heavier, with complicated latches, and more prone to leak. But no one had thought about a crew trying to escape a raging fire. That simply could not happen, and for the first 6 years of manned spaceflight it didn't, not until January 27, 1967.

Horrible as the fire was, it certainly changed things for the better at Downey and the other space places. A new seriousness of purpose, a sense of dedication to detail, pervaded. The clean room crew, behaving less like jesters and more like acolytes, greeted us astronauts with solemnity, as survivors of a holocaust.

After some finger pointing on each side, NASA and North American got on with the business of analyzing the fire and making sure it could not recur. Every precaution was taken to prevent sparking, and to thwart the spread of flames should a spark occur. A painstaking materials survey eliminated most conventional fabrics and substituted some exotic newcomers. The nylon outer layer of the pressure suit, for example, was replaced with Beta cloth, a woven glass fiber. But Beta's wear resistance was very poor, and in zero gravity tiny particles of the fiber could flake off and float through the cabin, and be inhaled by the crew and lodge in their lungs. The Beta cloth had to be combined with something to make it more abrasion resistant. Back to the laboratory. Finally, a thin coating of Teflon was added, and the new suits could survive brief exposure to temperatures as high as 5,000°F.

One by one, combustible components were replaced with fire-resistant materials. The one exception was paper: in the form of checklists and emergency procedures books, it was essential and no satisfactory substitute could be found. But henceforth books would be stored in protective bags or compartments.

The various changes had to be selected, approved, codified, implemented, documented, and inspected. Each manager had his own ways of getting things done. When someone got stubborn Joe Shea would take the offender outside into the parking lot and point out the moon. "Want to get there?" he'd ask. That usually did it.

George Low's specialty was the meeting of the group known as the Configuration Control Board, which gave final approval to any and all design changes. In Low's description, "Arguments sometimes got pretty hot. . . . In the end I would decide, usually on the spot, always explaining my decision openly and in front of those who liked it the least. To me, this was the true test of a decision—to look straight into the eyes of the person most affected by it, knowing full well that months later, on the morning of a flight, I would look into the eyes of the men whose lives would depend on that decision."

No item was too small to escape scrutiny. Joe Shea recalls that before the fire, his managers had kept a list of approximately 8,000 failures or irregularities that they were following. One item on this list, an order to North American to remove some flammable material from the lower equipment bay, arrived at Cape Kennedy the day of the fire. Had the order been implemented, Joe thinks, the fire would not have taken place. One piece of paper away from salvation, one radio headset away from burning up himself, today Joe Shea still chokes a bit at the recollection. Looking beyond the fire, however, he is emphatic: "I would not trade those years for anything, even though I get blamed for the fire."

In the 2 years after the fire, George Low's Board considered 1,697 changes and approved 1,341 of them. Many of these were fire-related, while others reflected various sins of omission or commission and improvements deemed worthy of the time, weight, and cost. Max Faget eased the flammability problem considerably by suggesting that, on the launch pad, 100% oxygen be replaced by a mixture of 60% oxygen and 40% nitrogen. Then when the spacecraft ascended and the pressure bled off to 5 psi, pure oxygen would gradually replace the nitrogen. This 60/40 mixture was a compromise between flammability and physiology.

In addition to making people very, very careful, the fire had another hidden but salutary effect. It gave everyone not working on fire-related matters a breather, a period to catch up on their work. In the race for the moon, no one wanted his piece of the machine to be the laggard, *the one* to hold up the whole procession. Consequently no one wanted to admit being "the long pole in the tent'' as it was called, and managers were apt to fudge their schedules a bit, hoping someone else would admit to being even farther behind. Many long poles got whittled down to manageable size during the time North American was struggling to get the Command Module back on track.

Not only did Joe Shea get eased out of his job, but North American reshuffled several top managers as well. The new space division president, Bill Bergen, put great emphasis on the position of spacecraft manager, the person held responsible for moving a spacecraft through its final stages of assembly and test. The spacecraft manager assigned to Command Module No. 101, the first of the post-fire Block II models, was John Healey.

John managed to transfer his own sense of urgency to his crew without trading quality for quickness. I asked Healey how he did this. ''I don't have sloppy in my vocabulary, but there are ways to speed things up.'' Once a worker wanted to tie up No. 101 for several days while he bonded some straps to the floor. Healey instructed him to go off and practice his technique somewhere else and come back when he could do it in a couple of hours.

Healey was a *production* man, not a development engineer. He wanted a clear line of demarcation between development and production. When he first saw No. 101, crawling around inside it were engineers, not workers! This incensed him. ''Never listen to the mouth; always look at the hands.'' He wanted production people in there, working with their hands, not engineers tinkering and talking about improvements. ''Engineering is like diarrhea,'' he told me disgustedly, ''it just keeps dribbling on forever.'' To put a stop to it, Healey denied the engineers access to No. 101. He told Bergen, ''Your phone is going to ring a lot,'' and it did, with various complaints about the autocratic Healey. But No. 101 began to look like a flying machine.

Still, with all the changes, and the requirement to test them exhaustively, it was not until May 1968—15 months after the fire—that No. 101 was delivered to the Cape. To the credit of John and his crew and their hours of overtime, it arrived with fewer discrepancies, according to the eagle-eyed Cape inspectors, than any of its predecessors. ''We care enough to send the very best,'' the Downey crew had carefully lettered on its shipping crate.

By October, No. 101 was ready for launch. Its flight was known as Apollo 7, under a numbering system that considered Grissom's crew to be Apollo 1 and that counted five more intervening unmanned tests. Its crew was Wally Schirra, Donn Eisele, and Walt Cunningham. Wally became the only man to fly on all three programs—Mercury, Gemini, and Apollo. Eisele and Cunningham were, like me, members of the third group of astronauts, and were on their first—and as it turned out, their only—spaceflight.

Apollo 7 was an 11-day flight and accomplished all its test objectives plus a few extras. Everyone who had anything to do with the CSM breathed a sigh of relief, because the machine had performed flawlessly. The fuel cells produced potable water, the SPS motor at the tail end of the Service Module got a real workout, and live television was beamed to earth. No one sweated No. 101's flight any more than John Healey. Normally not a worrier (his wife of 42 years insists he'll never get an ulcer), he today admits to having been a "little nervous." After the launch everyone left their consoles at the Cape to attend a celebration party in town, but John lingered on at his for hours, watching the data streaming down from No. 101, making sure his baby was healthy. It was, and the moon seemed a lot closer, thanks to No. 101.

The Lunar Module, on the other hand, was not yet ready for flight. For several months George Low had considered how to keep the program's momentum going while waiting for the lagging Lunar Module to catch up. The moon was the target—why not fly the CSM to the moon, or around the moon, in a scouting expedition preliminary to the landing?

Low asked Chris Kraft, Director of Flight Operations at Houston, to look into the idea. He also floated the notion with Bob Gilruth and Deke Slayton, Chief of the Astronaut Office. Finding no opposition in Houston, Low next approached von Braun in Huntsville. The second manned Apollo flight had originally been scheduled on a Saturn 1B booster, incapable of reaching the moon, but had been switched over to the Saturn V in an earlier effort to streamline the program. Von Braun was therefore being asked to adjust his thinking a second time in regard to his oversized offspring.

The third flight of the Saturn V, if Low had his way, would have changed swiftly from being an unmanned shot to a manned flight in earth orbit to a manned expedition to the moon; despite the fact that its predecessor had experienced trouble in all three of its stages. To Low's delight, von Braun agreed, saying that once the decision had been made to put three men on the Saturn V they might as well keep going all the way to the moon.

Jim Webb and George Mueller were more difficult to convince, but Webb agreed to keep the lunar option open for Apollo 8 provided Apollo 7 was an unqualified success.

With Apollo 7 behind him, George Low found all of NASA rallying behind a lunar Apollo 8 as a way of expediting the flight sequence. The Soviets did their bit to help by sending three unmanned Zond spacecraft to the moon. Big enough to carry a man, would the next Zond do exactly that and upstage Apollo? The decision to go for a Apollo lunar flight was not unanimous, however. Some knowledgeable people, like McDonnell's Walter Burke, thought the spacecraft should swing past the moon but not brake into lunar orbit. Years later, Joe Shea startled me by opening a conversation with, "Apollo 8 horrified me." Like many others, Joe thought that if you went all the way to the moon you should be prepared to land on it.

I had more than a passing interest in the flight of Apollo 8. Along with Frank Borman and Bill Anders, I had been assigned to the third manned Apollo flight, but spinal surgery in the summer of 1968 caused me to be replaced on the crew by Jim Lovell. When the first Lunar Module delivery schedule slipped, the crew assigned to it slipped also, and Borman, Lovell, and Anders leapfrogged into position to fly Apollo 8. Instead of flying to the moon, then, my job was to sit in Mission Control and act as CAPCOM, relaying advice and instructions to my old crewmates. At the time I thought that might be my closest contact with a lunar mission, and while I wasn't too happy about such a prospect, I figured it was better to be usefully employed as CAPCOM than sitting at home thumbing through the flight plan.

When Apollo 8 launched, on December 20, 1968, I was in Mission Control in Houston, hunched over my communications console and staring at a screen on the wall. On it was a diagonal green line, depicting the ideal path for 503 as it ascended from launch pad to orbit. My job was to watch as a second line, the radar track of the rocket, began to appear and superimpose itself (we hoped) on the first. If the second line started to deviate from the first, quick measurements were to be taken and, if necessary, I was to inform the crew to abort and fire the escape rocket that would blast the Command Module free. Fortunately Apollo 8 slid up that line like a squirrel hugging a tree trunk, and continued all the way to the moon along a near-perfect trajectory. George Mueller's all-up testing, George Low's imaginative leadership, and Wernher von Braun's confidence in his giant had all combined to give the United States a much-needed shove toward a landing by the end of the decade—just a year away.

Apollo 8 also gave the country a psychological as well as technical lift, as the crew spent Christmas Eve circling the moon. The first men to see the far side of the moon marked the event by reading from the Bible, the first 10 verses of Genesis, and concluding their television broadcast with " . . . Merry Christmas and God bless all of you—all of you on the good earth." For me at least, feeling sorry for myself in Mission Control, it was powerful stuff. Again the CSM performed brilliantly, including the SPS engine that burned for 4 minutes to slow Borman and company down to lunar orbital velocity and for another 3½ minutes to boost them back onto an earth-return trajectory. With two near-perfect performances by the CSM, it was now up to the Lunar Module.

The Command Module and the Lunar Module were a study in opposites. The Command Module was sleek and sturdy, the Lunar Module ungainly and fragile. The difference was due to the earth's atmosphere. The Command Module came screaming back from the moon at a speed of 25,000 mph and its tremendous kinetic energy had to be absorbed within the atmosphere. The Command Module structure had to be rugged to withstand the deceleration, its surface silky smooth because of the slipstream, its heatshield dense and sturdy enough to protect against a 5,000°F shock wave.

The Lunar Module, on the other hand, flew only in the vacuum of space, far above any annoying wind blast. It was launched inside a container and never returned to earth. An awkward-looking box with thrusters and antennas stuck on at odd angles without a thought being given to streamlining, the Lunar Module was also a delicate machine whose pressure shell could be pierced by a screwdriver dropped by a workman. Jim McDivitt called his Lunar Module Spider, and it was an apt name. The Lunar Module was built for the moon's gravity, one sixth of earth's, and its landing gear was so flimsy that on earth a fully fueled Lunar Module would collapse of its own weight.

The Lunar Module was built in two pieces, a descent stage that remained on the surface of the moon, acting as a launch pad for the ascent stage perched atop it. Over 70% of its total weight was fuel. Each stage was powered by a single rocket engine, which worried people a great deal. The descent engine was not quite so critical because if it failed during a landing attempt the crew could abort using the ascent engine in a technique called "fire in the hole." But the solitary ascent engine, like the SPS engine, simply had to work.

Both the ascent and descent engines used hypergolic fuels that ignited on contact, requiring no spark plugs, and both used bottled helium to pressure feed the propellants, eliminating the need for complicated pumps. Max Faget had insisted on these features. The Lunar Module was made as simple and reliable as possible by using proven concepts and well-tested machinery. Its designers even bypassed fuel cells in favor of batteries. It was a short-duration, special purpose machine designed for a very specific set of circumstances, thanks to the lunar orbit rendezvous decision.

The Lunar Module was built by the Grumman Aircraft Engineering Corporation at Bethpage, Long Island. The company had been established during the depression by Leroy Grumman, who had been a naval aviator during World War I, and although the firm had grown exponentially during World War II, in 1962 it was still a relatively small, tight-knit organization, and employees had a feeling of "family" there.

Ed Hiller is typical of those who worked on the Lunar Module. Like his father before him, Ed put in 25 years with the company. Once he considered quitting to go into an entrepreneurial venture with a friend, but the company talked him out of it. "For the first time in my life I felt I belonged," he told me. "It's the people, a long line of good smart people that make Grumman."

The Grumman people included a lot of pilots, some of them racing pilots, and Grumman planes reflected their experience. The company calendar for 1943 proclaimed, "The pilot never rides alone. With him . . . go the mind and heart and energy of all those who had wrought the miracle of mechanism which gives him mastery over gravity and space. . . . The plane he flies is not merely a thing of organic and inorganic material . . . the plane is a creature of man's genius, endowed with motion. Each fitting, part, and rivet; every fairing, calculated stress, and computed force is integrated into a machine that comes to life. . . . We work together with the Grumman star ascendant. . . . "

During World War II the Grumman star ascended through a series of stubby, sturdy Navy fighters, the best known being the F-4F Wildcat and the F-6F Hellcat, which produced more aces than any other U.S. fighter plane. Today our fleet's long-range fighter is the F-14 Tomcat. Just as on the West Coast

North American had specialized in Air Force planes, notably the highly successful P-51 Mustang of World War II vintage and the F-86 Sabre of Korean War fame, so had Grumman become almost exclusively a Navy supplier. Its planes were known to be solid, strong, even *heavy*, all exemplary adjectives but not ones ideally suited to describe the paper-thin Lunar Module. Weight would be a problem, a big problem.

One of the major advantages of LOR was that the landing craft could be stripped of all other duties and built for that one task alone—and could therefore be light. One of the Langley Center's early concepts was of an extremely simple vehicle, a 3,200-pound space scooter ridden by one man in a pressure suit. In the wake of the decision to land two men on the moon Houston produced more conservatively designed two-stage vehicles with enclosed cabins. One early Lunar Module looked like a badminton shuttlecock; another like a four-eyed bug. Grumman's original design was based on information supplied by NASA, and consisted of a cylindrical descent stage with five landing gear, topped by a more or less cylindrical ascent stage with two docking ports and four windows. Unlike the Mercury, Gemini, and Apollo Command Module shapes, the design of the Lunar Module cannot be traced to one person. "It evolved," says Max Faget. Caldwell Johnson and Owen Maynard of NASA and Tom Kelly of Grumman were probably the closest to being the Spider's parents.

NASA expected the Lunar Module to weigh 25,000 pounds, and Grumman's 1962 design was a couple of thousand pounds lighter than that, but then the inevitable growth ensued as new requirements were added and as actual subsystems and equipment turned out to be heavier than planned. By 1965 the Lunar Module topped 30,000 pounds and when it finally flew to the moon it weighed in at slightly over 33,000 pounds.

Adding weight created a vicious cycle: more weight meant more fuel required

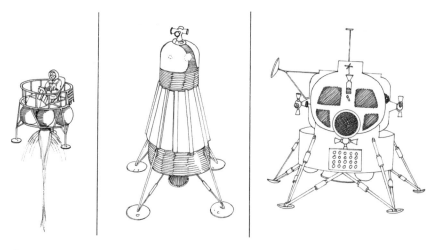

Space scooter   Shuttlecock   4-eyed bug

to lift that weight; more fuel, bigger tanks; bigger tanks weighed more, and so on. At the time of Grumman's contract award, many components were not well understood. Stark Draper's guidance system, for instance, "grew substantially" according to Joseph G. Gavin, who was Grumman's vice president in charge of the Lunar Module. Joe, a tall, spare Bostonian who had not lost his accent despite having worked for Grumman on Long Island for 40 years, remembers fighting weight every way he knew how: "a daily struggle."

He brought in outside experts. He formed an SWIP team (Super Weight Improvement Program) that second-guessed the entire design, poring over blueprints to shave a pound here, an ounce there. Four windows were replaced by two, canted inward and close to the crew's eyes for maximum visibility. The front docking hatch was simplified and made into a hinged door. The most radical change was to remove the crew seats. The astronauts agreed they could fly perfectly well standing up, and their legs could easily absorb the slight shock of touching down in $\frac{1}{6}$ G. All that was needed was a simple system of straps to hold themselves in place, although at one time bar stools were considered.

Grumman had decided on five landing gears because that configuration was very stable and well suited to cope with the irregularities in the lunar surface. But as the Lunar Module grew it became apparent that to fit inside its cavity atop the Saturn V, the gear would have to fold—another complication. Also, the structure of the descent stage was cruciform, or "tic-tac-toe" as Joe Gavin described it. These factors combined to cause Grumman to simplify the design, eliminating one leg and attaching the remaining four to the cardinal points of the frame.

"We agonized over the landing gear," Joe Gavin recalls. Joe had lunch with Tommy Gold, the scientist predicting a thick layer of lunar dust, and emerged incredulous: "He assured me of 10 meters of dust." Clearly Grumman couldn't design to those extreme conditions—the Lunar Module would be buried! But what *was* reasonable? Ranger, Surveyor, and Lunar Orbiter photos would not be available in time to assist in this decision. Between them NASA and Grumman arbitrarily selected footpads 1 yard in diameter, separated (diagonally) by 30 feet. The landing impact was absorbed by struts filled with a crushable honeycomb material, avoiding the more common hydraulic shock absorbers and their complexity.

Joe put his computer people to work, simulating hundreds of lunar landings under the worst possible conditions. They tried the equivalent of landing on a sheet of ice and skidding into a curb; hitting on four inclined planes, which would cause the Lunar Module to spin; one-, two-, three-, and four-pothole cases; front leg down, rear leg up; etc. They decided that a sink rate at touchdown of 7 mph might occur, and Joe ended up building struts that could compress 28 inches. "They never used more than two or three," he told me. The conservatism of the Navy aircraft designer? "Don't be *too* conservative, but make it work—that's my philosophy," Joe replied.

Work the Lunar Module did, and beautifully. But in the process of changing, refining, simulating, and fighting weight, Grumman slipped farther and farther behind schedule. To begin with, they started a year behind North American, thanks to NASA's struggles with the lunar orbit rendezvous decision. Had it not been for Jim Webb, Grumman would have been even farther behind. Joe Shea recalls a luncheon in November 1962 with the President's scientific advisory committee, a distinguished group headed by Jerome B. Wiesner and including a Nobel Laureate. Wiesner was against LOR and had assembled his troops to so inform Webb, for about the tenth time. "Jim," he said, "I know you are not a technical man and cannot understand all this, but I want you to hear from my experts." Webb listened, and when the experts were finished replied, "Jerry, you've held up this program long enough. Unless I hear from the President, I'm going to let the Lunar Module contract." Wiesner and his group were silent.

James Webb

In the aftermath of the Command Module fire in January 1967, Grumman got a slight breather, but the fire itself imposed another group of modifications. Add an ounce here, shave an ounce there. In addition, the ascent engine had serious problems. The descent engine should have been the troublemaker, as it was the first throttle controlled rocket engine to be used in a spacecraft. By injecting varying amounts of inert helium gas into the flow of propellants the thrust level could be altered without changing flow rates. It was a clever idea but a tricky one, and complicated the development of the engine.

The ascent stage engine, on the other hand, was a straightforward constant-thrust design. The problem was combustion instability. It kept flunking bomb tests: when a small charge was intentionally exploded inside the combustion chamber, the engine could not regain normal operation, and structural damage followed. As in the case of the Saturn V's first stage F-1 engine, the problem was traced to the fuel injector, but the engine's manufacturer, Bell Aerosystems, could not fix it. NASA hired Rocketdyne, makers of the F-1, to work on an alternate injector, although George Low worried about changing horses in midstream. After much soul-searching, George decided to ship the Bell engines to Rocketdyne for assembly with the Rocketdyne injector.

Although the ascent engine was the most serious Lunar Module problem, there were an assortment of lesser headaches. During a test a window blew out of Lunar Module No. 5, which was probably going to be the first to land on the moon. The descent stage developed oxidizer leaks. Cracks were found in the aluminum structure of several Lunar Modules and broken wiring was a nagging problem. NASA was pressing Grumman to hurry up and fix these problems.

Gavin was trying to comply but at the same time, like John Healey at North American, he wanted to DO THINGS RIGHT. It was his credo. "We argued like hell when we thought we were right." Comparing it to the airplane business, Joe added: "We understood the limited number of Apollo missions, on worldwide TV, with crews at risk. We couldn't fly around the field, and there were no ejection seats or parachutes. We had to DO IT RIGHT."

By early 1968, Lunar Module No. 1 was finally ready to try an unmanned flight. This was primarily an engine test with three firings of the descent engine and two of the ascent engine, including one "fire-in-the-hole," a simulated abort maneuver in which the ascent stage blasted free of the descent stage. The tests successfully completed, Lunar Module No. 1 entered the atmosphere and, unprotected by a heatshield, caught fire and plunged into the Pacific Ocean near Guam. The flight was so successful that George Mueller decided a second unmanned flight was unnecessary, and Lunar Module No. 2, which had been earmarked for that purpose, was later donated to the National Air and Space Museum, where it is still on exhibit.

With two manned CSM flights (Apollo 7 and 8) plus the Lunar Module No. 1 experience, it was time to put the two machines together in earth orbit. This 10-day flight, Apollo 9, fell to Jim McDivitt, Dave Scott, and Rusty Schweickart, and consisted of a complex and demanding series of tests. The key element was putting two men into the Lunar Module, separating them from the CSM, and checking out all the components of the rendezvous system.

McDivitt and Schweickart, in Spider, backed away from Scott in Gumdrop until they were over 100 miles behind, and then slowly returned, methodically checking all their gear, especially their rendezvous radar. The process took over 6 hours and culminated in a successful docking. Of course, if the rendezvous had failed, only Scott would have lived to tell about it, because the Lunar Module had no heatshield and could not survive a plunge into the atmosphere. Rusty Schweickart, despite bouts of nausea, conducted a brief foray out onto the Lunar Module's "front porch," testing out the backpack that would be used on the moon. The only significant abnormality was some roughness in the descent engine caused by uneven helium ingestion.

I asked Schweickart whether he felt uncomfortable 100 miles away from Dave Scott, who had the only heatshield in the sky that day. "Not at all," he replied, "I felt safer than in some of the altitude chamber tests in Houston . . . the Lunar Module was a lot more responsive to the controls than the CSM, like a fighter plane instead of a transport. It was also very comfortable standing up in the Lunar Module compared to the CSM, which I always felt was cluttered and confining." To each his own. I'll take a lot of clutter for one solid heatshield.

The next flight could have been the first lunar landing, but NASA decided to stick to its plans and conduct a dress rehearsal. The first three manned flights had been unqualified successes, but there were still many unanswered questions, such as: was the descent engine's throttle responsive enough for a landing?

Also, the landing radar had not yet been tested in flight. The Lunar Module had never been to the moon, and things were different there.

During rendezvous, for example, lighting conditions, orbital velocities, and ground tracking capabilities were all considerably different than in earth orbit. Furthermore, we had discovered that the gravitational pull of the moon was not evenly distributed. There were heavy spots called mascons (mass concentrations) that would cause the Lunar Module to dip slightly as it passed over them. How would this affect rendezvous? A practice rendezvous, one in which the Lunar Module did not land but stayed in the same orbital plane as the CSM, might be a sensible precursor to the more complicated situation of a lunar takeoff as a starting point for rendezvous.

But the clincher was Lunar Module No. 4. Despite the best efforts of Joe Gavin's Super Weight Improvement Program, Lunar Module No. 4 was overweight, fatter than Lunar Module No. 5. Perhaps some fuel could be off-loaded, but fuel margins of safety were difficult to predict, and what with all the other unknowns—mascons, landing radar, helium ingestion—the overweight Lunar Module became the straw that broke the camel's back. So the next flight, Apollo 10, would fly to the moon and the Lunar Module would swoop down to within 50,000 feet of the landing site, but then "take a wave-off," as a carrier pilot would describe it, and climb back up to the CSM.

The crew of Apollo 10 were all veterans of a couple of Gemini rendezvous: Tom Stafford, Commander; John Young, who would stay in the CSM Charlie Brown; Gene Cernan to accompany Stafford in Lunar Module No. 4, Snoopy. They departed earth at noontime on May 18, 1969. Their Saturn V gave them a rough ride, with Pogo oscillations in both first and second stages, and a growling, rumbling, vibrating third stage. Despite that, however, it put them on such an accurate trajectory that they required only one tiny midcourse correction, adding 30 mph to their velocity.

When they got to the moon it looked to Stafford like a big plaster of paris cast. They paid particularly close attention to the proposed Apollo 11 landing site and the area looked to them smooth enough for a landing. They called the approach path to the landing zone "U.S. 1" and they called out distinctive features along this imaginary highway. Their rendezvous turned out to be routine except for one exciting moment when Snoopy gyrated wildly, the result of a control switch in the wrong position. "Son of a bitch!" Cernan yelled on the radio, for all the world to hear, before he and Stafford regained control of the situation.

Like McDivitt before him, Stafford found the near-empty ascent stage, at less than 6,000 pounds, a totally different flying machine than when the two stages were full of fuel and weighing 33,000 pounds. The fully loaded Lunar Module was sluggish, but when the same thrusters were used to control the near-empty ascent stage, they whipped it around like a race car. With the crew safely back in Charlie Brown, ground controllers fired the ascent stage engine

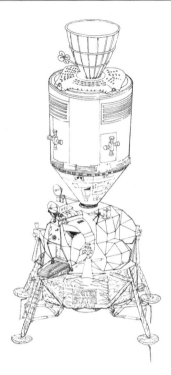

CSM and the Lunar Module docked

to fuel depletion and sent it into an orbit around the sun. On May 26, Apollo 10 plopped into the Pacific near Samoa and the stage was set for the main event.

When I finished my stint as Apollo 8's CAPCOM, I was assigned, along with Neil Armstrong and Buzz Aldrin, to the crew of Apollo 11. From my earlier crew assignments, I had become a CSM specialist, and I would be in charge of what we later named Columbia, alias Command Module No. 107. Neil was the commander and he and Buzz would pilot Eagle, Lunar Module No. 5, and attempt the first landing, at the western edge of the Sea of Tranquillity. If you look at the full moon, this spot is slightly right of center. More precisely it is located 23° east of a vertical centerline and 1° north of a horizontal center-line. It was picked more for piloting reasons—a smooth landing field—than for scientist interest in its topographical features.

As a former astronaut I am most frequently asked whether I wasn't terribly disappointed to go all the way to the moon but not walk on it. It is certainly true that I did not have the best seat on Apollo 11, and I would be a liar or a fool to say otherwise. But it is also true that I am perfectly satisfied with the seat I did occupy. I had always considered the first landing flight to be the premier one of the series, and I was proud to be a part of it—any part of it. By the time I joined Armstrong and Aldrin it had been clear to me for over a year that I had become a CSM specialist, for reasons beyond my control, and it was equally clear to me that of the three of us, I was the logical choice to stay in Columbia.

So I had no quarrel with either of my crewmates. I respected and admired them both.

Neil was the premier test pilot of the astronaut group, and Buzz the most learned, but that didn't mean that Neil was all pragmatist or Buzz just a theoretician. Among X-15 pilots Neil was considered to have had the most complete understanding of the machine's complex systems. Buzz's credentials included not only a doctoral dissertation on orbital rendezvous but some success as a college pole-vaulter and a Korean War MiG killer. As a pair, they brought to the flight a formidable array of talent, and I felt very fortunate—especially with my recently repaired neck—to be joining them. I tended, and still do, to compare myself not with the two of them, but with the scores of equally qualified astronauts and the millions of people everywhere who would have gone in my place. I was lucky, and I knew it.

Early in 1969 the three of us started 6 months of intensive preparation to ensure that Command Module No. 107, Lunar Module No. 5, and the three of us would be ready by July to do what President Kennedy had asked 8 years earlier. Not only did the crew and the flying machines have to be readied, but also the entire NASA ground apparatus—the tracking network, the mission control center, even a quarantine facility to house a crew that might bring lunar bugs back with them.

We astronauts knew how dependent we were on ground equipment. Nowhere was this more apparent than in the case of the simulators. They taught us to fly, these incredible machines with subtleties and complexities that even their flying counterparts did not possess. Their job was to duplicate, insofar as they could, the spacecraft and the space environment. Zero gravity was beyond them; on earth weightlessness can only be achieved for half a minute or so by flying a high performance aircraft in a parabolic arc, or by free-falling, and that is not practical. Launch and entry acceleration could be practiced on a centrifuge, but most of the Apollo simulators were "fixed based"—that is, they just sat there motionless, in 1 G. Past this one limitation, however, they were truly marvelous.

"The great train wreck," John Young called the Command Module simulator, consisting of what appeared to be a jumbled pile of boxes with a staircase leading up to an exact copy of the Command Module side hatch. The "boxes" were actually a complex arrangement of scenes and mirrors to simulate passing and oncoming views of landscapes, lunarscapes, etc. Inside the hatch was replicated the interior of the Command Module: same size, same shape, same colors, same equipment. Only a real expert could tell the difference between the interior of the simulator and that of an actual spacecraft. And the instruments all worked! An imaginary oxygen leak could be detected by a drop in tank pressure; an overvoltage condition would transfer equipment from one electrical bus to another.

All these dials, controls, switches, and so forth were hooked up to a room full of computers that had stored in them the flight characteristics of their respective subsystems, and that fed up-to-date information back to the cockpit. One or more instructors sat at a console equipped with instruments to monitor our performance. Out the windows the crew could see earth, moon, sky, and stars. Here the fidelity broke down, and the visual effects were crude, but for training purposes they were more than adequate. It didn't matter that in the simulator the star Antares was not precisely its true shade of red, only that it be in the proper position for a measurement with the sextant.

In the months before a launch, we astronauts practically lived in the simulators. For Apollo 11, I spent over 400 hours training in the Command Module simulator alone, usually by myself, but frequently with Neil and Buzz. On special occasions, Neil and Buzz would be in the Lunar Module simulator, I in the Command Module simulator, and we would be hooked up to Mission Control. Together we would run through the delicate phases of our flight over and over again. To complicate the situation, simulated "failures" could be inflicted on us by a team of specialists. All the way to the moon and back we would be tormented by a series of riddles. Some problems, such as arcane electrical malfunctions, could be solved only after considerable discussion among ourselves and perusal of the extensive library of schematic drawings we carried with us. Others, such as an errant rocket motor, caused our vehicles to veer

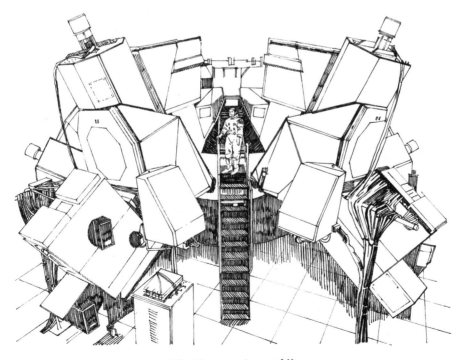

The "great train wreck"

wildly out of control and had to be countered within seconds by memorized procedures. As we became more expert, our instructors became more sadistic.

While the three of us were honing our skills, so were the Mission Control teams. There were four of them, identified by colors, and they worked in shifts. Each specialized in a particular part of the flight (launch, rendezvous, lunar landing, entry) and each contained specialists in the various subsystems of our machines. Mission Control had been at the Cape during Mercury days, but had moved to a fancy building in Houston early in the Gemini program. It was a model of hierarchical decision-making, a pyramid with outside experts at the bottom, specialists in the middle, their supervisors toward the top, and at the very apex, the Flight Director.

"Flight," as the Director was called, ran both a democracy and a dictatorship. The process was democratic in that he actively solicited the views of his constituents and gave great weight to their counsel, but he tolerated no uncontrolled babble of voices. He had to make the final decision, frequently within seconds, and the pyramid had strict rules and procedures to assist him in the process.

Information bubbled up from below, synthesized and condensed as it rose. Questions were fired back down the line and descended as far as necessary to produce satisfactory answers. Some experts were located in anterooms and talked by phone with their bosses who were in a large central chamber whose front wall was crammed with electronic maps and supplemental information. Facing this were tidy rows of consoles, workstations featuring cathode ray tubes on which could be displayed thousands of bits of information transmitted by the spacecraft and crew. In the center of this temple of technology sat "Flight," the high priest, and near him his pet astronaut, CAPCOM, who relayed his distilled wisdom to the crew.

It was a remarkably efficient system and in all the years of manned space flight I know of no case in which Mission Control has given bad advice to a spacecraft in flight. Sometimes during the simulations, yes, but never during an actual flight. The original "Flight" was Christopher Columbus Kraft, Jr., and when Chris began training others to replace him, he picked only the finest— people in his own mold, that is, smart, strong, and opinionated.

During Mercury and Gemini days, communications had been constantly interrupted as the orbiting spacecraft passed out of view of the nearest tracking station. NASA had put a station on every strategically located island (Ascension, Madagascar, etc.) and supplemented these with two tracking ships, but still, there was mostly water in the equatorial regions beneath the spacecraft, and each page of the crew's flight plan was dotted with annotations reading AOS and LOS: acquisition of signal and loss of signal.

It was a nuisance cramming all talk and exchange of information into an AOS zone. For example, when Neil Armstrong and Dave Scott had tumbled out of control on Gemini 8, they were LOS, out over the Pacific, and Mission Control was unaware of their problem and unable to help them isolate the thruster that had stuck open.

Going to the moon was much easier, because in cislunar space the high-flying spacecraft was always in view of whichever half of the earth was turned in its direction. Hence no AOS-LOS zones, but continuous communications. The three primary tracking stations for lunar missions were located near Madrid, Spain; Honeysuckle Creek, Australia; and Goldstone, California. Spaced evenly around the globe, at least one of the three was always in view between earth and moon. Each featured a huge, 85-foot diameter parabolic antenna, and even a quarter of a million miles away we enjoyed high-fi, static-free communications. In fact, they were so good and so continuous that sometimes I was tempted to turn the radio off just to enjoy a little quiet.

Behind the moon it *was* quiet—very quiet. Each lunar orbit lasted 2 hours, and for 48 minutes of that period the spacecraft was out of sight of the earth and, therefore, unable to talk to it. The interplay between Mission Control and the spacecraft (and consequently the decision-making process) varied considerably, depending on whether communications were sporadic, continuous, or nonexistent. As in the case of Gemini 8, the crew always had to be prepared to take matters quickly into its own hands, but generally speaking, the crew deferred to Mission Control because not only was there more collective wisdom there, but the ground's information was usually more complete and more precise.

For example, I was trained to navigate back from the moon by using my sextant to measure the angles between selected stars and the earth's horizon; however, there was no way I could determine our position as accurately as the giant earth-based tracking radars, and the only point in my training was to take over in the event of lost communications. Similarly, during launch the crew depended almost entirely on the ground for information about the health and happiness of the Saturn V.

On the other hand, docking was a pilot-controlled, manual process, with Mission Control a helpless bystander. Likewise, the final phase of the lunar landing was almost totally under crew control. But generally speaking, data was telemetered in a constant stream from the spacecraft back to earth through one of the three big antennas, and from there sent to Mission Control in Houston, where it was processed by four IBM 7094 computers and delivered to the base of the pyramid, there to be analyzed and filtered for "Flight's" inspection and possible action. The people in Mission Control flew every mile of the way with us.

One last ground facility would be used only in the event of a successful landing and return. As early as 1962, Carl Sagan and other scientists were worrying about lunar microorganisms being brought back to earth where they might multiply explosively. Lunar rocks, the returning spacecraft, and the crew members themselves needed to be decontaminated. The government responded by forming the Interagency Committee on Back Contamination, which examined the process NASA proposed for recovering the crew and spacecraft and quarantining them. Some within NASA did not take this work too seriously because they thought the moon was a sterile, lifeless place. What would happen, they asked, if they just used the same old Mercury and Gemini recovery procedure? In one meeting the representative of the Public Health Service replied, "emphatically," that in such a case the crew would not be allowed back in the country! I guess, like the man without a country, they would have been condemned to ply the high seas for the rest of their contaminated lives.

At any rate, NASA built the Lunar Receiving Laboratory in Houston to house rocks, craft, and crew for an observation period of 21 days beginning at the time of contamination, the lunar landing. In other words, we would be locked up for around 17 days if all went well. If it didn't, who knew? Neil tried to explain our situation in mathematical terms: if you multiplied a very small number (the probability of our being contaminated) by a very large number (the consequences to the planet of such contamination), what size answer did you get? We didn't know, but apparently it was a finite number large enough to be reckoned with. Hence the Lunar Receiving Laboratory, where we would live behind a biological barrier along with a minimum support staff plus a large colony of white mice. The mice, exposed to us and our rocks, had better stay healthy, or we might be in the Lunar Receiving Laboratory for a long, long time.

Possibilities such as these did not worry me particularly. Once we returned to earth I figured all would be well; it was the catastrophes along the way that I feared. Suppose, for example, Neil and Buzz were stranded on the moon? Their ascent stage engine wouldn't start, and there they were, with a day's supply of oxygen. How would NASA handle *that*? Would NASA pull the plug or keep broadcasting their final words to the world? What would *I* say or do?

Years later my curiosity about such matters overcame me and I asked Julian Scheer, Jim Webb's public affairs chief, how he would have handled the situation. Was there a plan for death in the file, just as there seemed to be a plan for everything else?

"Yes, there was a plan," he told me. "I was going to uphold the public's right to *know*, but not a right to see or hear. Just like we would never release the tape of the Grissom fire; if we had two astronauts stranded, Mission Control would continue to communicate in private with them. The public has a right to know that there has been an accident and how bad it is, but they don't have to witness or hear it."

Would he hand out suicide pills before a flight? "Just like Gary Francis Powers was supposed to carry them in the U-2? We kicked the idea around, but not seriously. If a crew requested them, that would be their option, but none of the astronauts took that idea seriously."

Nor did I. But perhaps more than most astronauts, I was impressed by the complexity of our machinery and the acceptance of the fact that one or more parts of it might break. I am not a good mechanic and if my car fails to start

A fragile daisy chain

or my toilet to flush, I need help. To me, Apollo 11 was a ½-million mile daisy chain draped around the moon, a chain that was as fragile as it was long. I figured our chances for a successful landing and return were not much better than fifty-fifty, despite all the help we had gotten from our pathfinder friends.

Our chances of returning with our skins, however, were far better than even, especially mine, because I did not have to depend on the Lunar Module. From all I knew the Lunar Module was a most reliable machine, but still . . . there were just so *many* things that could go wrong.

Shortly before the flight of Apollo 8, Jerry Lederer, NASA's Safety Chief, had summed up my own view. While the flight posed fewer unknowns than had Columbus' voyage, Jerry said, the mission would "involve risks of great magnitude and probably risks that have not been foreseen. Apollo 8 has 5,600,000 parts. . . . Even if all functioned with 99.9% reliability, we could expect 5,600 defects. . . . ''

Of all the links in the daisy chain, the one that I fretted most about was the rendezvous. Perhaps because of our excessive fuel usage on Gemini 10, I was almost morbidly aware of how swiftly a rendezvous could turn sour. A tilted gyro, a stubborn computer, a pilot's error—ah, it was that last one that troubled me the most. If Neil and Buzz limped up into a lopsided orbit, would I have enough fuel and enough moxie to catch them?

The basic rendezvous was fairly simple but I would carry with me a notebook listing 18 variations on the theme, and some of them required quick, deft, flawless performance on my part. For example, normally I would be in a 60-mile circular orbit as Neil and Buzz launched at a precisely calculated time and shot for a 45-mile orbit, our height difference of 15 miles selected to give them a leisurely catch-up rate. But suppose, for any of a host of reasons, they were late taking off. Then, instead of a 45-mile orbit they would aim lower because, according to Sir Isaac Newton's laws, the lower an orbit, the faster the speed, and they could make up for lost time and catch me before they ran out of oxygen.

With no atmosphere surrounding the moon, as they got later and later they could reduce the height of their orbit until they were barely skimming the lunar mountaintops. But what if they were later still? Then the strategy suddenly was reversed and the hunter became the hunted. Neil and Buzz would then go as high as they could, into a slower orbit, and I would dive for the hills, speed up, and make an extra turn around the moon to catch them. That was just one of the 18 cases in the rendezvous book that I would have clipped to the front of my pressure suit. Of such possibilities are nightmares bred.

By launch day, July 16, 1969, I found myself with tics in both eyes, tense about rendezvous and a hundred other things, but nonetheless eager to abandon the tedium of the simulator and reenter the exciting world of spaceflight—nearly 3 years to the day after Gemini 10.

I was everlastingly grateful that I had flown in space before. The Gemini 10 flight received worldwide attention, but still it had a local or at most a national flavor, and it was treated in a friendly, uncritical way—almost as if it were a sporting event. This Apollo 11 flight, on the other hand, was serious business, with the ghost of John F. Kennedy riding with us in full view of a watchful and wary world. It had taken nearly 10 years and the work of 400,000 Americans to bring Apollo 11 to launch.

Superimposed on the normal crew pressure—the nagging worry of "what have I overlooked?"—was the imperative *not* to fail, to represent all the experts of this country, to distill their knowledge into 8 flawless days, to drape that daisy chain around the moon and return it undamaged to earth. Not only was I worried about my tiny pink body, but I had the typical pilot's preoccupation with the possibility of a mission failure attributed to his error. I think that the pilot's prayer is to be found in the 25th Psalm: "My God . . . let me not be ashamed. . . ." The accident records are full of cases of pilots killing themselves to avoid being ashamed; not taking remedial action soon enough because they

Landing sites of Apollo missions to the moon

didn't want to admit their foul-ups. One of my proudest recollections of Apollo 11 was that the three of us did our jobs well.

Six more Apollo flights to the moon followed in the next 2½ years. Their paths to and from the moon were carbon copies of Apollo 11, but each visited a different spot, selected for its geological attractiveness, and each expanded in some way our capability to explore. They stayed longer, they ventured farther from the Lunar Module, they brought back a heavier load of rocks, they deployed more complex scientific instruments. The final three carried a lunar roving vehicle, a lightweight, battery-powered dune buggy that enabled them to traverse over 20 miles of lunar surface. The final flight, Apollo 17, stayed on the moon over 3 days and brought back over five times as heavy a cargo of rocks as had Apollo 11 (257 pounds versus 46 pounds).

If things had worked out a tiny bit differently, Apollo 12 might have been the first landing flight and Pete Conrad the first man on the moon instead of the third. Missing by such a slight margin might have soured some people, but not the ebullient Conrad. He acknowledged his place in history, and his 5'7" frame, as he jumped from the last rung of the ladder onto the lunar surface.

"Whoopee! Man, that may have been a small step for Neil, but that's a long one for me!"

On his third spaceflight, Conrad and rookie Al Bean laughed and cavorted as they explored the surface, while Dick Gordon had my job overhead in his Command Module, Yankee Clipper. While Armstrong could put down anywhere he found the terrain suitable, Pete had an extra complication. He was to guide Intrepid to a spot next to Surveyor III, which nearly 3 years earlier had landed in Oceanus Procellarum, the Ocean of Storms.

As usual, Pete did his job well. His instructions were to stay at least 500 feet away from Surveyor, so that he wouldn't blow dust on it, and he ended up 600 feet from it. Also different from Apollo 11 was the array of scientific instruments Conrad and Bean erected on the surface. A solar wind measurement device, seismometer, magnetometer, ion detector, laser ranging reflector, and several other instruments constituted ALSEP—the Apollo lunar surface experiment package. An ALSEP was emplaced at all subsequent Apollo landing sites, and returned useful data for years thereafter.

Apollo 13 never made it to the surface of the moon. Outward bound, 200,000 miles from home, its crew heard something we all dreaded: a loud bang. Jim Lovell, Jack Swigert, and Fred Haise watched in horror as both Service Module oxygen tanks emptied.

An explosion in one tank had also caused a leak in the other, and NASA's putative redundancy evaporated just as quickly as the oxygen cloud they could see streaming out behind them.

"Hey, we've got a problem here," they reported to Houston in a masterpiece of understatement.

Fortunately they still had their Lunar Module, and they used it as a lifeboat, hoarding its limited supply of oxygen and electricity. Of course they had to hang onto the moribund Command Module because they needed its heatshield for protection during entry and its parachutes for landing. The temperature inside dipped to 38°F, water condensed on the walls, and sleep became almost impossible. The crew was "cold as frogs in a frozen pond," according to Jim Lovell, but they hung on for nearly 4 days, limping home just barely before their oxygen ran out.

Apollo 14 saw good fortune return to Apollo for keeps. Al Shepard, the first American in space in 1961, landed on the moon in 1971, the long delay due to his being grounded with an inner ear ailment. His companion in the Fra Mauro hills was Ed Mitchell, a Navy captain who believed in extrasensory perception and who attempted, unsuccessfully, to transmit numbers mentally to a friend back on earth.

Al and Ed used a rickshaw-like device to haul scientific equipment, and operated a seismic experiment called "the thumper," in which 13 small charges were detonated along a straight line and their reverberations measured by geophones. Al Shepard also hit a golf ball. Stu Roosa operated a precision camera in the Command Module while awaiting their return.

Shepard's analysis of the first few missions: "Neil, Buzz, and Mike proved that man could get to the moon and do useful scientific work, once he was there. . . . Apollo 12 and 14 proved that scientists could select a target area and define a series of objectives, and that man could get there with precision and carry out the objectives with relative ease and a very high degree of success."

Apollo 15 made me feel a little bit old because for the first time one of our spaceflights was commanded by someone younger than me. Dave Scott, Jim Irwin, and Al Worden were all Air Force officers and they named their landing craft Falcon (the mascot of the Air Force Academy, although Scott and Worden were West Point graduates, and Irwin an alumnus of Annapolis).

Scott and Irwin, landing near Hadley Rille at the foot of the Apennine Mountains, were the first to carry the jeep-like dune buggy called the lunar roving vehicle. Shepard and Mitchell had gotten a good workout walking 2 miles in a 4½-hour stint while Scott and Irwin rode over a dozen miles in comfort. Weighing less than 500 pounds, the battery-powered Rover was an ingenious machine. Even its birth amazed me. The astronaut pulled a lanyard on the side of the Lunar Module, and, like a newborn giraffe, the Rover plopped to the ground and unfolded. Each of its four wheels had its own electric motor. and

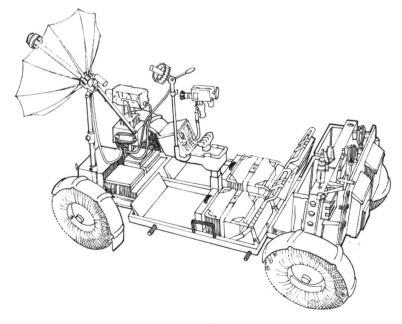

Lunar Rover

the wheels themselves were made of wire mesh. It could do about 11 mph, downhill.

Scott and Irwin spent a total of 18 hours on the surface, collecting samples and conducting experiments. Dave even had time for a gravity demonstration. Holding a geologist's hammer in one hand and a falcon feather in the other, he dropped both simultaneously, and cameraman Jim Irwin recorded the fact that they hit the ground at the same instant, there being no atmosphere to slow the descent of the feather.

Even in $\frac{1}{6}$ G astronauts found working on the lunar surface fairly strenuous because of the bulky pressure suits and the equipment they had to manipulate using clumsy gloves. After one of Jim Irwin's forays, biomedical data from sensors taped to his chest indicated that his heart was behaving a bit strangely, skipping some beats. In later years, Jim suffered a series of heart attacks. At the time, doctors thought that perhaps his diet had caused a chemical imbalance, and more potassium was added to the menu of the next flight.

Apollo 16 visited the light-colored highlands of the moon, adjacent to the crater Descartes, a region believed to have the most ancient rocks available for collection. It was rolling, hilly country, with plenty of big boulders, and was about 8,000 feet higher in elevation than Tranquillity Base. My old friend John Young was the Commander, accompanied by Charlie Duke on the surface and Ken Mattingly overhead.

John seemed to like everything about this flight except the flatulence caused by the potassium-rich orange drink that had been urged on him. When he complained about it to Duke, as luck would have it a microphone switch stuck open and John inadvertently broadcast this epic report from the moon: "I mean, I haven't eaten this much citrus fruit in 20 years. But I'll tell you one thing—in another 12 fucking days, I ain't never eating any more. And if they offer to serve me potassium with breakfast, I'm going to throw up. I like an occasional orange, I really do, but I'll be damned if I'm going to be buried in oranges."

Apollo 17 ended the proud series. The Lunar Module Challenger departed the moon on December 14, 1972, to rendezvous with America. No one has been back since. Originally there were to be three more Apollo flights, but one by one they were canceled as the risks seemed to outweigh the returns. Of course the geologists wanted every nook and cranny of the moon examined, but the public and NASA's managers seemed to have lost their enthusiasm for it.

The crew of Apollo 17 was a little bit different in that it included a scientist for the first time. Joe Engle, a good old fighter pilot, had been scheduled to accompany Gene Cernan to the surface, but pressure from the scientific community caused him to be replaced by Harrison H. ("Jack") Schmitt, a Harvard-trained geologist who had worked for the U.S. Geological Survey in New Mexico, Montana, and Alaska. Ron Evans occupied the Command Module. Schmitt's name is more familiar than most because from 1977 to 1983 he was the U.S. Senator from New Mexico.

He and Gene Cernan, in their Rover, prowled a valley of the moon known as Taurus-Littrow, looking for evidence of volcanism. They found no volcanoes, but they did bring back rocks covering a 2-3 billion-year stretch of the moon's history, and I'm sure that the sharp eye of the experienced geologist increased the selectivity, and hence the scientific value, of the specimens collected. They left a plaque on the leg of their Lander similar in design to the one left at Tranquillity Base by Neil and Buzz, but with a slightly different message: "Here man completed his first exploration of the moon, December 1972 A.D. May the spirit of peace in which we came be reflected in the lives of all mankind." When Apollo 17 splashed down in the Pacific it had logged 12½ days, the longest of the Apollo flights.

Despite the failure of Apollo 13, and even considering the deaths of Gus Grissom, Ed White, and Roger Chaffee, the Apollo program was more successful than I thought it could possibly be back in October 1963 when I became an astronaut. To me the moon was so far away then, so beyond our reach or our grasp, that I thought a dozen preliminary manned Apollo flights would be required, instead of four. Of course, Gemini established a base of confidence in rendezvous and long-duration flight and a nucleus of experienced flight and ground personnel, so Apollo didn't really have to start from scratch, but still, to me it was a miraculously successful adventure.

At its peak, Project Apollo employed 400,000 Americans in government, industry, and universities. Twenty thousand industrial firms were involved in a large or a small way. It cost $24 billion, well within the range of estimates given President Kennedy ($20 to $40 billion). The fire cost three lives, 18 months, and $400 million. Still, 12 men walked on the moon, four before the end of the decade. Of all the technical challenges, the Saturn V was the one most likely to cause massive failures and delays, but from its first flight it worked almost to perfection. Not a single Saturn V, nor its predecessors, the Saturn I and IB, ever exploded—a record unmatched by any other family of rockets.

The Lunar Module turned out to be not as fragile as it appeared, and its successes included, on Apollo 13, saving its sturdier sister, the CSM. One fire inside a Command Module, one tank explosion inside a Service Module: far from perfection, but much better than I thought possible in the early sixties, as we planned the quantum jump from Mercury to Apollo. An entire Mercury capsule could fit easily inside an F-1's engine bell. Nor was it simply a matter of making things bigger, they were much more complex as well. Consider the blind plummet to earth of a Mercury after retrofire, compared to the precision of an Apollo Command Module navigating home from the moon, slicing into the atmosphere at exactly the right angle, dodging thunderstorms, and plopping down next to the carrier deck.

A million factors contributed to Apollo's technical success. In my view, most notable among them were the idea of clustering five rocket engines together, the use of hydrogen as fuel, the experience of von Braun and his team, the lunar orbit rendezvous scheme, not invented by John Houbolt but championed by him, Max Faget's genius, Bob Gilruth's steady hand at the helm in Houston, George Mueller's all-up testing, and Grumman's standard of excellence.

Chiseled into the facade of Union Station in Washington, not far from where President Kennedy pointed us toward the moon, is one of my father's favorite quotations: "He who would bring back the wealth of the Indies, must carry the wealth of the Indies with him." After the flight of Apollo 11, Neil, Buzz, and I had the honor of addressing a joint session of Congress, and in my remarks I repeated these words, saying that we had taken to the moon the wealth of this nation, the vision of its political leaders, the intelligence of its scientists, the dedication of its engineers, the careful craftsmanship of its workers, and the enthusiastic support of its people. In return, we brought back rocks. In those rocks, I went on, was locked the mystery of the origin of the moon, and indeed even the origin of our earth and solar system.

Today, nearly 20 years later, to my great disappointment, that secret has yet to be unlocked. The rocks we brought back from the Sea of Tranquillity are mostly basalt: dark, fine-grained igneous rocks formed by the cooling of molten lava some 3.7 billion years ago. Later flights to the lunar highlands returned with older, lighter colored igneous rocks called gabbros and anorthosites. These lunar rocks have answered many questions but, as with most advances in science, several new puzzles have emerged from every one solved.

Before geologists had examined any moon rocks, there were three main theories of the moon's origin. The first suggests that the moon condensed from gaseous matter just as the earth did, at approximately the same time and in approximately its present place. The second is that the moon was a stranger wandering through this solar system that was captured by the earth's gravity. Third is that the moon used to be a part of the earth, perhaps of the Pacific basin, that broke loose. The lunar samples make the first and third theories difficult to accept because the moon rocks have a different chemical composition than do earth rocks. The second theory is vulnerable to mathematical analyses that indicate it is highly unlikely that a body the size of the earth could attract and capture in its orbit something as large as the moon. Thus the rocks we brought back have failed to answer one of mankind's most ancient questions—where did the moon come from? Frustrated scientists work on. Perhaps, they say now, a huge object collided with the earth 4 billion years ago, and the moon was formed from the resulting debris.

Beyond the technological success of Apollo, it was certainly a victory for the human spirit, if your spirit is one that values exploration. Mankind's most complex expedition brought back to earth not only moon rocks but a renewed sense of confidence in the ability of the United States to succeed at whatever it deemed to be in its national interest.

In the years since, that feeling has diffused and perhaps evaporated entirely in the wake of the shuttle Challenger explosion. Yet national projects still seem to be evaluated on the basis of "If we can put a man on the moon, why can't we . . . ?"

Clearly there is a long list of things we cannot yet do, such as cure cancer. And another long list of things we won't do, such as stopping the slaughter on our highways. But Apollo has given us some guidelines for success.

First, its objective was clearly and starkly defined, without any possibility of misunderstanding: to land a man on the moon and return him safely to earth by the end of the decade. Second, all available resources were put to the task. Jim Webb was brilliant in building the widest possible base with industry, academe, and the government working harmoniously together. Third, contingency planning—asking over and over "What happens if?"—was a vital part of the program, and an invaluable one when an inflight emergency occurred, such as on Apollo 13. Fourth, the team members were encouraged to state their views, as forcibly as they wished, before a choice was made, but to pitch in and fully support the decision afterwards. The selection of lunar orbit rendezvous is a good example of this process.

Apollo's spinoffs, or benefits to other parts of our economy, have been debated endlessly. Apollo has been given credit for everything from microelectronics to the Teflon frying pan. The argument usually hinges on what might have been done with Apollo money had it been spent elsewhere, a highly speculative exercise that makes it difficult to make valid comparisons. Certainly satellites have worked their way into our lives in the fields of worldwide communications, weather prediction, and military reconnaissance. Beyond that, the industrial base spawned by Apollo is still struggling to establish itself on a firm commercial footing in other fields, such as the manufacture, in space, of exceptionally pure pharmaceutical products or oversized crystals for the electronics industry. Use of space as a platform for observation has also produced interesting results, as the next chapter of this book will show.

I do recall one clear, positive aftereffect of the first lunar landing. Traveling around the world several months after the flight, I was continually impressed by the fact that no matter where we were, the reaction was the same and, to me, unexpected.

Never did I hear, "Well, you Americans finally did it." Always it was "we," we human beings drawn together for one fleeting moment watching two of us walk that alien surface. Too bad that feeling could not be sustained, that we cannot suspend our terrestrial squabbles as we did one time, on July 20, 1969.

It is said that on that night a note appeared on John F. Kennedy's grave in Arlington. Attached to a small bouquet of flowers, it read: "Mr. President, the Eagle has landed." And with 5 months to spare before the decade's end.

5

Life in Orbit

While Apollo 11 was still returning from the moon on July 22, 1969, NASA Headquarters informed its Centers to proceed with the dry workshop program. What was a "dry workshop," and how did it gain approval?

President Kennedy's decision to go to the moon had leapfrogged the traditional sequence of space exploration. Tsiolkovsky, Oberth, and many others wrote about earth-orbiting space stations, considering them both a vantage point for looking back at the earth and a staging location for expeditions beyond.

Wernher von Braun, writing in *Collier's* magazine in 1952, described in some detail a space station in the form of a gigantic wheel, rotating to produce artificial gravity. Just a year after Sputnik, in 1958, a committee of the House of Representatives, in a report entitled *The Next Ten Years in Space*, concluded that a space station was the next logical step. Industry was also interested, and the press. In 1960 the *London Daily Mail* chose "A Home in Space" as the theme for its annual Ideal Home Exhibition. Douglas Aircraft submitted the winning proposal, a space laboratory built into the empty upper stage of a hypothetical launch vehicle, and 150,000 people walked through a mock-up of it.

Within NASA, the Langley Center took the lead in space station studies in the early 1960s, and work of this type continued later in Houston and Huntsville. At NASA Headquarters, George Mueller wanted a program to proceed in parallel with Apollo, something to maintain momentum in case Apollo faltered, or to extend beyond it if Apollo enjoyed early success. Also, some scientists were criticizing Apollo as being a stunt, or at best a mere demonstration of technology, and devoid of scientific interest. Using Apollo hardware in earth orbit to conduct scientific investigations would allay this type of criticism. Furthermore, in

Huntsville, von Braun was fretting about the future of his organization once the Apollo program drew to a close.

Therefore it was only natural that NASA examine the idea of an earth-orbiting habitat in exhaustive detail. To use existing hardware to the maximum extent possible, the choice boiled down to one Saturn V and one Saturn IB, or two or more of the smaller Saturn IBs. In the former, or ''dry,'' scheme the third stage of the Saturn V (the S-IVB) would be outfitted on the ground as a laboratory, and launched empty. The Saturn V could reach earth orbit without requiring the energy in the third stage. So it could be launched dry and be ready for the crew when it arrived from its Saturn IB ride.

The ''wet'' plan involved launching a fully fueled Saturn IB and then conducting a rendezvous with it by three men in another Saturn IB. The crew would vent and purge any remaining hydrogen from the fuel tank of the near-empty second stage (again an S-IVB), bring various items of equipment into it, and conduct experiments there. This approach could be expanded by a cluster concept, using three or more Saturn IBs. Whether launched wet or dry, the S-IVB, with its huge interior volume, nearly 10,000 cubic feet, was an attractive starting point for considering a comfortable and commodious habitat about the size of a three-bedroom house.

As time went by, Huntsville and Houston became competitors, Huntsville favoring wet and Houston dry. Also, for a while Houston insisted that zero gravity was an undesirable quality and that a space station had to be rotated fast enough to produce at least one-tenth of the earth's gravity. Gradually, however, Houston relented, and equally gradually it became apparent that the dry concept was far simpler and would result in a much more fully equipped and safer laboratory and habitat module. *If* a Saturn V could be spared.

With Apollo 11's lunar landing safely over, NASA Headquarters sent the message that earmarked a Saturn V for a dry workshop. First known as the Apollo Applications Program, the dry workshop program was renamed Project Skylab, and resulted in nine astronauts on three crews spending 28, 59, and 84 days in earth orbit. It was the next logical step as NASA saw it, one that kept the ball rolling using as much Apollo hardware as possible, but one that would not subvert a longer-range goal that might emerge, such as a Mars mission. Quite the contrary: if one wanted to go to Mars, first the men and machines would have to be tested closer to home—aboard Skylab, or something like it.

The Douglas S-IVB stage was, as space machinery goes, quite simple: two tanks and a motor. For Skylab, the motor was not needed, and other equipment could be installed in its place. The hydrogen tank was over three times larger than the oxygen tank, and therefore the logical choice for the habitat. The smaller oxygen tank was used as a gigantic trash basket.

Inside the hydrogen tank, which was slightly over 20 feet in diameter and nearly 50 feet high, the space was divided into two floors separated by a metal grid. The upper, the larger of the two, was called the forward compartment, a cavernous space ringed by storage lockers. The lower floor contained the standard divisions of an earth-based house, more or less. There were three tiny bedrooms, a bathroom, an all-purpose living room/dining room/kitchen called the wardroom, and a work area.

Attached to the top of the forward compartment was an airlock module, in turn connected to a docking adapter. When the astronauts were in residence, the Apollo CSM would be plugged into the docking adapter. With a lot of practice and a little luck, it was possible for an astronaut to push off from the floor to the work area and fly up the length of his domain, 30 yards' worth, through the forward compartment, the airlock, the docking adapter, and into the Command Module—all without touching the walls.

The trick, Pete Conrad told me, was to put on a little Kentucky windage to allow for air currents along the way. Pete, the commander of the first Skylab mission, claimed to have scored at least one victory, going the whole way without brushing against the sides. I asked his crewmate Joe Kerwin about this, and he reacted more like a fighter pilot than a medical doctor (Joe is both). "Bullshit!" he cried. "Undocumented! That's like a guy going out on a golf course by himself and claiming he shot a hole in one." Anyway, compared to previous spacecraft, Skylab offered a huge interior volume, one that allowed not only fun and games but a serious look at living in space for a couple of months at a time.

The environment of space is in many ways safer and more predictable than that of earth, and it is certainly more quiet and serene, but there are some hazards as well. Just as a great white shark may devour a bather once in a decade, so may a meteoroid rip through a spacecraft. Radiation is always present, trapped by the earth's magnetic field in two belts named after James Van Allen, the University of Iowa physicist. Manned spacecraft generally fly below the inner Van Allen belt, but it dips down in the vicinity of the South Atlantic and an astronaut's exposure to it must be continually monitored.

Solar flares send waves of high-energy particles streaming from the sun. Here on earth our atmosphere screens them, but in orbit an eye must be kept peeled for unusual solar activity. High-energy cosmic rays come from beyond our galaxy. Then there is the vacuum on the other side of the pressure hull, a fraction of an inch away. If a window blows out, the cabin decompresses explosively, the crew loses consciousness in about 20 seconds, and death follows within a few minutes. Weightlessness, enjoyable as it is, gradually weakens the body to the extent that it may not be able to survive a return to earth's gravity.

Spacecraft designers cope with these problems as best they can. For Skylab, a meteoroid shield was added to the outside of the S-IVB. Launched flush against the rocket's skin, once in orbit it popped out some 5 inches. It was designed to cause a meteoroid to break into harmless fragments before it hit the S-IVB, and it also acted as a sunshade to keep the interior comfortably cool despite a surface temperature of approximately 250°F in direct sunlight. The crew was provided with dosimeters that constantly recorded radiation levels. If the dosimeters indicated that the total radiation absorbed by a crew member had reached a precalculated danger level, then the mission would be ended. Warnings of solar flares, which might cause a sudden dosage increase, would come from ground observatories as well as from an array of five solar telescopes aboard Skylab.

If the orbital workshop decompressed suddenly, the plan was for the crew to retreat to the Command Module and slam the hatch shut. If the break in the pressure hull was very small, a quarter inch or so, the crew had a repair kit that could plug the hole. The crew's physical and mental well-being would be checked in a thousand ways, and it was no coincidence that Dr. Joe Kerwin was a member of the first crew. The plan was to double and redouble the flight duration, building on the 14 days of Frank Borman and Jim Lovell aboard Gemini 7.

Considerable study was given to the composition and pressure of the gases to fill Skylab. A pressure of 5 psi was selected, continuing the tradition—and light weight—of Mercury, Gemini, and Apollo. One unexpected consequence of the reduced pressure was that sound wasn't transmitted nearly as well as on earth. The crew learned early that—even shouting—their voices didn't carry and they were forced to use the intercom if they were more than 10 feet apart. Instead of using 100% oxygen, a mixture of 70% oxygen and 30% nitrogen was picked. Thus the lungs would inhale about the same number of oxygen molecules as in ordinary air at sea level, yet Skylab's susceptibility to fire would be substantially reduced from that of its predecessors.

Outfitting the workshop was no small task. With three crews of three men each expected to occupy it for months, there were nearly 12,000 items to be catalogued and stowed aboard the 200,000-pound station. Food alone—over a thousand meals' worth—was a major problem in packaging and storing. In addition to carrying the food and other expendables required by previous spacecraft, Skylab had to be equipped as a radio and TV station, a physics and biology laboratory, a doctor's office, a photo studio, and an astronomical observatory.

Keeping track of 12,000 items requires organization and discipline. On earth, gravity keeps things in place. Aboard Skylab, opening the wrong compartment door could cause a 10-minute chase, as dozens of loose objects tumbled out and floated away in what Houston engineers dubbed the "Jack-in-the-box effect." Eventually, lost articles could be found sucked up against a filter screen covering the intake to the ventilation system, located in the top of the forward compartment. But the whole process of inventory and control of loose items was a damned nuisance, one that occupied and irritated the crew mightily.

Fortunately, the empty oxygen tank located below the workshop, and connected to it by an airlock, was available to be used as an oversized trash container into which empty food packages and other detritus could be thrown and forgotten. With no washing machine on board, the crew simply threw their dirty laundry into the airlock.

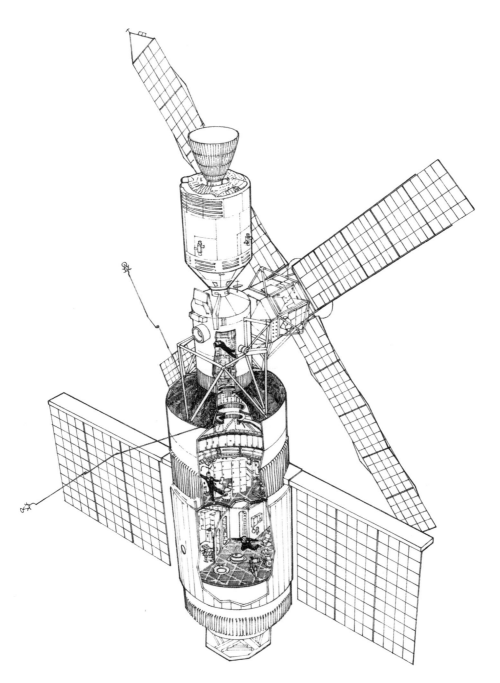

Inside Skylab

Above the oxygen tank was the huge hydrogen tank, the central compartment of Skylab known as the orbital workshop. Its lower portion, called the aft compartment, looked somewhat like a small apartment on earth, except that it was circular and the rooms wedge-shaped. Its floor and ceiling consisted of a metallic grid into which the astronauts could anchor themselves by inserting and twisting triangular cleats on the soles of their shoes. The wardroom was

Shoe restraints

where the crew gathered to eat and talk and look out their circular window, 1½ feet in diameter, at the earth.

Even after months, most of them never tired of looking at the earth, which they regarded as "down," "up," or "out there," depending on how they aligned their bodies. If they approached the window feet first, which they rarely did, the earth appeared "down," below their feet. Conversely, if they were head first and had to crane their necks, the earth seemed "up." Usually their bodies felt more comfortable parallel to the window, and then the earth was simply "out there." Inside the relatively cramped Command Module, I usually felt that the earth, for some inexplicable reason, was slightly down and to my left. Below what, or to the left of what, I do not know.

The matter of up and down was of more importance to Skylab crews than to their predecessors because of Skylab's larger volume and the variety of its workstations. The workshop itself was designed as if it were still on the launch pad. Lettering on the walls, controls, and gauges, etc., were all right-side up if the astronaut's feet were pointed "down" toward the oxygen tank and his head "up" toward the airlock module, the docking adapter, and the Command Module.

Within the docking adapter, a different system was used, with equipment arranged 360° around the circular walls. The astronauts aligned their bodies with whatever piece of equipment they were operating. As Joe Kerwin put it,

"You carry with you your own body-oriented world . . . in which up is over your head, down is below your feet . . . and you take this world around with you wherever you go."

Still, the astronauts felt more comfortable when their own body axis was in alignment with the axis of the compartment they were in. To Caldwell Johnson, it was as simple as watching television at home, lying on your side. "If something interesting comes on, you sit upright." Walking on the wardroom ceiling, said one of the crew, was "a strange sensation. You see brand-new things . . . it's a fascinating new room." Occasionally in the docking adapter someone would look into the workshop and get an uncomfortable feeling that he was about to "fall down" into it. But in general all members of the three crews adapted easily and comfortably to their strange new home.

In the bathroom the floor was solid (for ease of cleaning) instead of having holes for shoe cleats, and some of the crew missed that system for anchoring themselves. Without it, one said, "you just ricochet off the wall like a BB in a tin can." The toilet used airflow as a substitute for gravity to draw feces and urine from the body and collect them in bags. Some samples of each were retained and frozen for post-flight medical analysis, and the rest was dumped into the oxygen tank. A collapsible, cylindrical shower enclosure was provided. Once

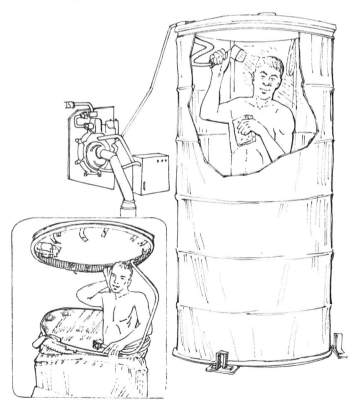

Skylab shower

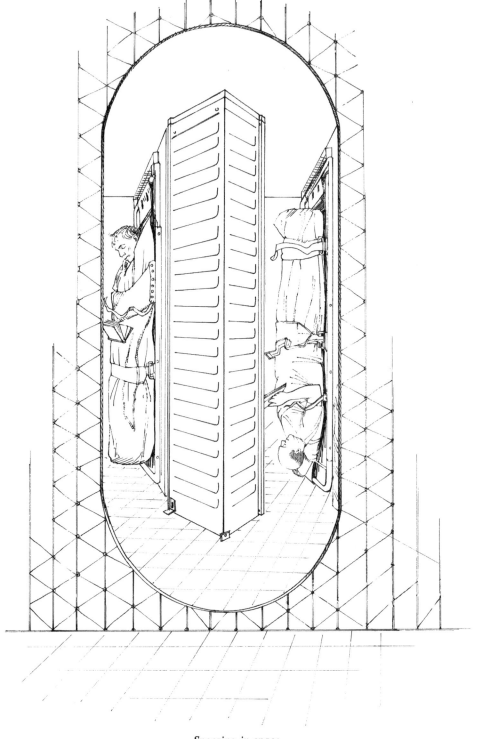

Snoozing in space

inside, the astronaut drew it up around him and sealed it closed. He squirted water on his body with a hand-held shower head, and again airflow was used instead of gravity to collect the waste water. Some astronauts preferred a rubdown using wet washcloths, of which they had over 800, color coded for each man's use.

The bedrooms were tiny compartments, just large enough for a thin sleeping bag attached to the side wall. In relation to the wardroom and bathroom the crew slept "standing up," although Pete Conrad didn't like the flow of ventilating air blowing up his nose in this position, and he reversed his bag and slept upside down. It didn't matter to him. Weightlessness is very comfortable for sleeping, with no pressure points on the body, just floating with the light touch of the sleeping bag to reassure the sleeper that he isn't straying off somewhere. In weightlessness the body must be forced to stretch out straight. When relaxed, the torso and knees bend slightly in a semi-crouch, and the arms float out in front, with elbows bent and fingers dangling in front of the face. Aboard Apollo, I missed having a pillow, and used to drift off to sleep with my head wedged up against the end of my sleeping bag.

Getting up in the dark to go to the bathroom could be an adventure. As Al Bean, the commander of the second crew, explained it to me, he could slip out of his sleeping compartment with no problem, but when he turned in the direction of the bathroom, his feet left the floor and suddenly he was floating in a different world, unable to visualize where the nearest light switch might be, now banging into something, maybe the ceiling or maybe a wall, he wasn't sure.

In addition to the wardroom, bathroom, and bedrooms, the aft compartment included a work area containing the trash airlock and the ergometer, a device that occupied a fair amount of the crew's time. Basically an instrumented stationary bicycle, the ergometer was the exercise machine that would assist each man in keeping fit by putting a daily load on his cardiovascular system, as well as his arm and leg muscles. Those who spent the least amount of time on the ergometer, like doctor Joe Kerwin, were in the worst shape on landing and required the longest time to regain their preflight health and strength levels. Trust a physician to ignore his own advice.

The upper level of the workshop was much larger than a designer would have provided, given a blank piece of paper. However, Skylab had started life as a hydrogen tank designed to boost Apollo to the moon—so there the giant was, over 20 feet in diameter and 27 feet high, separated from the lower level habitat by a grid work partition.

This cavernous upper chamber had been equipped with a blue aluminum fireman's pole, to allow the astronaut a safe traverse, but they found the pole interfered with their fun and they removed it. This forward compartment was their gymnasium, as well as their storehouse and experiments center, and they loved to play there, using their weightless freedom to carom off the walls in

Skylab crew acrobatics

ever more complicated flips, cartwheels, and arabesques. With just a little practice, they could do things that Olympic gymnasts on earth might envy, but never duplicate. They also bounced balls off the walls, threw darts, and flew paper airplanes, but they most enjoyed using their own bodies as if they were divers, suspended effortlessly between board and water. A triple gainer with a reverse twist? No problem!

In the center of the dome that formed the forward compartment's ceiling there was a circular hole leading into the airlock module. Here was conducted the business of keeping the space station going, via the control panels for the electrical system, the environmental control system, and the communications equipment. The airlock module was so named because it contained the port through which the astronauts exited to perform extravehicular activity. Nitrogen and oxygen were carried in six spheres and six cylinders outside the module, and the two gases were mixed to maintain a breathable atmosphere throughout Skylab. Carbon dioxide exhaled by the crew was removed by a molecular sieve, a more advanced system than the lithium hydroxide canisters used by Apollo and its predecessors.

Also unlike its predecessors, Skylab used the sun to provide electricity. Two immense solar panels were designed to extend out 35 feet on either side of the workshop, like stubby rectangular wings. Each contained over 30,000 solar cells about the size of a postage stamp, made of silicon and designed to convert the sun's radiation into an electrical charge. Altogether, one wing could produce over 2,000 watts of electricity. Since 30 minutes of each 93-minute orbit was in the earth's shadow, batteries were provided to store the electricity. In addition, there were myriads of other devices, such as circuit breakers, regulators, and switches, that were monitored by specialists in Mission Control.

Beyond the airlock module was the docking adapter, which performed several functions. As its name implies, it was where the Apollo Command Module plugged in, and it contained an extra docking port in case a second Command Module had to be sent up in a rescue mission. It was also a center for experiments involving the study of the earth and the sun. The earth was viewed using a wide variety of instruments. One consisted of six cameras with carefully matched lenses, bolted together to take simultaneous pictures of the same object, but each using different film and filter combinations so that the subject would be recorded across a wide spectral range, from the visible to the infrared.

Other instruments recorded the intensity of radiation given off by the earth's surface, and its radar reflectivity. Skylab's orbit was nearly 300 miles high and inclined to the earth's equator at an angle of 50°, a higher inclination than other American manned spacecraft. Out their window the crew could see as far north as Labrador and as far south as Tierra del Fuego. As it criss-crossed back and forth, Skylab overflew three fourths of the earth's surface and 90% of its population.

The study of the sun was by far the most complex experiment aboard Skylab, at least from the point of view of the equipment involved (nothing, I suppose, is more complex than the human body). When the Skylab was on the launch pad, it had attached to its nose a wondrous machine called the Apollo telescope mount (ATM), a name inherited from an earlier, canceled mission.

The heart of the ATM consisted of five instruments to record the sun's spectrum all the way from visible light to high-energy X-rays, a coordinated approach to solar research never before attempted. Once in orbit, the ATM was pivoted around to the side of the docking adapter, where its four solar panels gave the Skylab the appearance of a Dutch windmill. At least it looked that way to me if I ignored the rectangular wings protruding out below. Parts of the ATM could be operated by remote control from the ground, but in the main it was man-

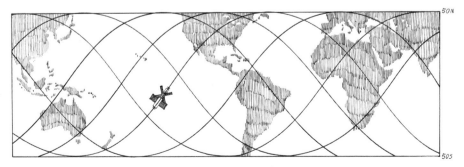

Skylab orbital path

ipulated by the crew using a very complicated control station inside the docking adapter. Exposed film was retrieved from the ATM periodically by extravehicular forays, a good reason for the crew to be pleased with the instrument, as they enjoyed getting outside for a change.

On the tip of the docking adapter was the port into which the Command Module (with its Service Module) was plugged. The CSM was standard Apollo equipment, although if a rescue had become necessary, some interior modifications to the Command Module would have been necessary to hold the extra crew. The docking port was normally left open so that the crew could beat a hasty retreat in the event of an emergency down below. The Command Module was also a quiet corner into which a crewman might retreat to read or look out the window.

With a CSM attached to Skylab, the entire pile of machinery was 130 feet long, with a wingspan at the solar panels of 90 feet. It weighed 100 tons and had an interior volume over 50 times that of an Apollo Command Module. It was a machine formidable but forgiving in its size and complexity, and one that, while certainly not luxurious, had been designed with crew comfort and convenience in mind. Like all flying machines, its design was the result of many compromises.

For example, the matter of the wardroom window was of paramount importance to both the crew and the design engineers—but for different reasons. The crew, pilots all, could not imagine zooming through that glorious orbital sky month after month, over every nook and cranny of the earth between 50° north and 50° south latitude—all that, sealed up like sardines in a can, with no way to look out of the workshop. Preposterous!

The Skylab director in Huntsville, on the other hand, felt that a window was a pleasant thing, a nice touch, but come on—this Skylab had to be designed for efficiency and the window was a frill *not required for mission success*! A window was too expensive, it would take too much time to design, build, and test, and it wasn't on the original S-IVB. But worst of all, a window offended an engineer's sense of structural integrity and design elegance. Here you had this great rounded cylinder, this symmetrical beauty of a hydrogen tank, and through the side of it, through an unnecessary *hole*, would protrude a *window*? It was a violation, a weakening of the structure, a rape almost.

One day in the fall of 1969, George Mueller convened a meeting in Washington to discuss issues of habitability aboard Skylab. The wardroom window was high on the list. Mueller had included Raymond Loewy, one of the country's best industrial designers and creator of the sleek 1953 Studebaker, and when the Huntsville troops circled their wagons, George turned to Raymond for advice. It was unthinkable, Loewy said, *not* to have a window, for recreation if nothing else—and thus the issue was decided.

In other cases, however, the crew thought more like engineers (which most of them were) than psychologists. In the early Apollo days, some consideration was given to the matter of up-down orientation, and someone suggested a paint scheme inside the Command Module to solve the problem. Below a certain line, the interior would be painted earth brown, above it sky blue.

We astronauts, accustomed to rolls and loops in our fighter planes, ridiculed the idea in favor of a simpler, monochromatic gray. On Skylab, Raymond Loewy preferred lemons and tans, but the crew didn't much care one way or the other. The final product was a conglomeration of brown, gray, gold, blue, yellow, and off-white—and again the crew seemed to ignore the issue, just as an engineer would. If an object functioned well, it didn't matter what color it was painted. At least that was the attitude of the astronauts with military pilot backgrounds. A couple of the scientist astronauts did complain inflight about the colors, mostly about their monotony and what they thought was too much brown.

Complicated as Skylab was, and given the higher priority of the lunar landing missions, it was not until May 1973, six months after Apollo 17, that the $2.5 billion program was ready for flight.

Skylab's Saturn V, the last one to fly, looked good at launch, but within a minute a near disaster occurred. The meteoroid shield was supposed to fit flush against the skin of the S-IVB, but the seal wasn't tight enough, and the intense wind blast squeezed into the gap and ripped the shield off. With it went one of the workshop's solar panels, and the other one was pinned in its stowed position by an errant piece of the shield.

With its meteoroid protection gone and neither solar wing available to produce electricity, Skylab was in big trouble. Oddly enough, the shield was not missed so much as protection against meteoroid collisions as it was for its thermal properties. Without the shield as a sunshade, the temperature inside the workshop gradually worked its way up to 150°F. Not only was this too hot for human habitation, but at this temperature medicine and food would begin to deteriorate, and even worse, the polyurethane foam lining of the S-IVB was expected to decompose and produce toxic gases such as carbon monoxide and hydrogen cyanide.

The loss of one solar panel and the inability of the other to deploy was also a crippling blow. Fortunately, the four ATM solar panels could be interconnected to the workshop's circuits, but they did not generate enough electricity to allow full functioning of all systems. The fuel cells of the Apollo Service Module would have to be pressed into service, but they would run out of fuel after 20 days, and then the crews would have to return to earth, far short of their 56-day goal.

NASA had planned to launch Pete Conrad, Paul Weitz, and Joe Kerwin the day after the workshop, but now their Saturn IB ride was put on hold while people tried to sort out the problem. The workshop was maneuvered so that the sun's rays hit it at a 45°angle, a compromise between the need to generate electricity and a fear that excessive heat inside would ruin the workshop's habitability. Unfortunately, keeping the workshop pointed in this direction produced some bizarre side effects. While the interior of the hydrogen tank was too hot, parts of the docking adapter were getting so chilled that controllers in Houston were worried about water pipes freezing. To prevent that they rolled the workshop to place the water loops in direct sunlight, a position that also produced more electricity. But workshop temperatures climbed past 150°F, and flight controllers had to juggle the huge machine back and forth, using up a lot of maneuvering gas in the process, but keeping it alive nonetheless.

In Houston and in Huntsville engineers were working around the clock to devise some sort of makeshift sunshade. There was a scientific airlock in the side of the workshop through which an object could be poked. The problem was that the airlock was only 8 inches square and an area of some 400 square feet had to be shaded from the sun. The design chosen, one of a dozen considered, was an elaborate parasol using rat-trap springs and telescoping aluminum rods. The canopy was made of a thin sheet of spacesuit material, a combination of nylon, mylar, and aluminum. Plans were also made for freeing the solar wing, using cable cutters installed on the end of a long pole.

Eleven days after the workshop had reached orbit, Conrad and company were sent after it, with their parasol, cable cutters, and Pete's boundless enthusiasm. Conrad, one of the second group of astronauts selected, was 42 at the time, a veteran of two Gemini flights and the third man to walk on the moon. Pete is a short, balding, compact man with a gap-toothed grin and hooked nose. He makes friends easily and laughs and jokes a lot. His credo was "If you can't be good, be colorful," but that is deceptive because Pete is far more than a colorful clown. He *is* good, in many ways: as an engineer, aviator, judge of people, communicator, and leader. Living in odd circumstances for a long period of time can cause problems to fester, but Pete's crew always seemed jovial, enthusiastic, and optimistic. It was hard to be grumpy around Pete; somehow, he wouldn't allow it.

I think a lot of his attitude rubbed off on Al Bean, the commander of the second Skylab crew, who had walked on the moon with Pete. Al could not quite match Pete's ebullience, but he had a stronger work ethic, and he constantly pushed Houston—in the nicest, friendliest way—to find more work for his crew. The third crew was much different.

When Conrad arrived at Skylab, he confirmed what Houston had suspected. One solar wing was completely gone, as was the meteoroid shield. The other wing was still furled, folded against the side of the workshop and held in place by a twisted strand of the meteorite shield. Conrad maneuvered the CSM into place and depressurized it. Weitz opened the side hatch and emerged part way, while Kerwin hung onto his legs. For an hour Weitz struggled with the wing, using all the tools he had available, but he couldn't dislodge it.

The next hitch was when Pete docked. The Command Module's probe would not engage Skylab's drogue. Finally, after running through several alternate procedures, the crew had to depressurize again, and partially disassemble the back side of the drogue, bypassing some of the electrical connections. It had been a long 22-hour day, a discouraging start to what they hoped would be a 28-day adventure.

The next morning when the crew entered Skylab, they found the interior of the workshop too hot to endure for long. It was a dry heat, "like the desert" to Weitz, like the boiler room on an aircraft carrier, thought Conrad. They had to spend most of their time in the docking adapter, which was much cooler, but they were able to poke their parasol out through the airlock and deploy it. It didn't spread out as evenly as hoped, but the temperature inside the workshop began a gradual fall to a tolerable level.

The electrical situation was still critical, however. They were using all the power the four ATM solar panels could provide, and several battery failures made prospects look dim for completing the entire mission as planned. After a lot of soul-searching, Houston decided to try an extravehicular foray to free the folded solar wing. Conrad and Kerwin went out and managed to cut the wing free, but its supporting beam still would not pop out perpendicular to the workshop. Tying a rope to it, the two of them hauled like stevedores. "I gave a mighty heave," reported Conrad, "whereupon everything went black and I shot up into the air . . . and by the time I settled down from my whifferdill back into the airlock area, the beam had come all the way up and was fully deployed." Skylab was saved.

The Skylab crews had two different ways of talking to the ground. The A channel was a direct radio broadcast, and all the world—or at least anyone listening on the frequency—could hear. The B channel was less direct; it involved talking into a tape recorder whose contents were "dumped" via telemetry when Skylab next flew over a suitable ground station.

Pete Conrad

The rules concerning use of the A channel were clear enough, but the B channel caused a major upset. Conrad had mistakenly told a preflight press conference that the news media would receive copies of the B transcriptions, when actually NASA planned to keep that information confidential. Inflight, Conrad himself asked for the chance, from time to time, to have a private conversation with the ground.

This seemingly inconsequential matter provoked violent reactions within NASA, with the public affairs side fearing loss of the Agency's credibility with the American public, and the medical doctors trying to maintain the sanctity of their confidential relationship with their patients, the astronauts. NASA management got whipsawed by the two sides, with their emotional as well as factual arguments, but in the end agreed to protect from public scrutiny the medical data, or any information the crew considered to be of an emergency nature. Even so, the use of tape recorders remained an item of confusion, and later, the third crew spoke words into their recorders that they regretted mightily.

The single most difficult adjustment for both the astronauts and the flight controllers was determining a suitable workload. Skylab was packed with experiments, and Mission Control was packed with experimenters who wanted more and more of the crew's time.

Each minute in orbit cost the taxpayers a bundle, and the astronauts were well aware that they had an extraterrestrial obligation to work hard. But *how* hard? During a 3-day Gemini flight, John Young and I had been glad to work 18-hour days. In the vicinity of the moon, certain events had to be scheduled without regard for anyone's circadian rhythm. But aboard Skylab, week after week, what was reasonable? Sixteen hours a day? One day off in seven?

Each of the three crews proved to be quite different. At the beginning Conrad and company felt overworked as they struggled to adapt to weightlessness. Tasks that had taken 15 minutes in the simulator stretched to an hour as they searched through a series of identical lockers for the right tools and then chased them around the workshop as they floated free. In the early days they were working well past dinner to complete their assignments.

Conrad felt things were disorganized and that he, Weitz, and Kerwin "were getting inefficient by rushing." He chided Mission Control, noting that there were "enough guys down there to think out the flight plan a little better than you're doing." Pete offered a list of suggestions, and as the three of them became more familiar with their surroundings, they gradually caught up and even had some free time to read and just look out the window. They decided not to ask for more work, however, to avoid setting unrealistically high standards for the two crews that would follow them.

The first crew had to spend a large portion of their time coping with the unexpected. First the high temperature in the workshop and the rigging of the sun shield, next the shortage of electrical power and the freeing of the solar panel. Then there was a series of minor repairs, and one time Conrad even took a hammer outside and banged on a battery charger until a stuck relay broke free. Despite their slow start, Pete's crew finished all of their planned medical experiments and most of their solar and earth observations. They took 29,000 pictures of the sun and nearly 10 miles of magnetic tape for earth resources experiments.

The pictures of the sun did not come as easily as they expected. The jackpot, as far as the solar scientists were concerned, was a flare, a gigantic outburst of energy from the sun's surface. The sun has an 11-year activity cycle, and Skylab flew during a period of relative solar calm, making it even more important for the crew to catch one of these rarities on film. The ATM console included a flare detection system but it was constantly tricked by the South Atlantic anomaly. As Skylab passed over the eastern part of South America, the inner

Van Allen belt frequently dipped low enough to trigger a false alarm, the flare klaxon would go off, and one of the crew would race for the ATM console only to discover he had been hoodwinked once again. It was especially exciting when the system went off while all three were sound asleep. Joe Kerwin, in particular, seemed to be the favorite victim, and after the flight he evaluated the detection system as "absolutely worthless."

Nonetheless, the crew's persistence paid off, and after 3 weeks of frustration they finally hit the jackpot. Paul Weitz happened to be at the console, doing some catch-up work on what was supposed to be a day off, when he detected unusual activity through one of his instruments. The solar scientists were overjoyed that Weitz managed to record the entire sequence of the flare's rise and fall, playing his console like a pipe organ, as he focused the right telescope and film combination on the right spot on the sun.

By the fourth week, Pete told me, he was more than ready to come home. He and the others were fascinated by what they could do in weightlessness and what they could see out the window, but 28 days was what they had contracted for, and enough was enough. In general they gave Skylab's designers high marks for the workshop's habitability, but although it was a comfortable hotel, its restaurant was terrible. They were hungry and wolfed down plenty of calories, but for some unexplained reason the same food in space didn't taste as good as it did on the ground. Food seemed to lose its flavor, as if they had left their taste buds back on earth, and they suggested that the later crews carry spicier foods, and sauces and seasonings to liven up the dishes they had found especially bland.

More important even than their diet was their exercise routine. They had set aside a half-hour each day to ride the ergometer, but frequently something interfered, and Kerwin especially could not seem to find time for it. Part of the problem was that the bicycle-type seat was not well suited to weightlessness, even though they strapped themselves down. Conrad made the most use of it, sometimes pedaling with his arms instead of his legs. When they returned to earth, it was possible to correlate exercise time with physical condition. Joe Kerwin was noticeably wobbly, while Pete Conrad was in better shape than after his moon trip and Paul Weitz was somewhere in between. They recommended that the later crews spend more time exercising, which they did.

Al Bean's crew was scheduled for 56 days (twice Pete's), although they ended up doing 59. Al Bean and I were compatriots in the third group of astronauts, and I can't think of anyone easier to be cooped up with inside Skylab for 2 months than Al, except for one thing: his work ethic is more strongly developed than mine. He is an inquisitive soul who probes to find out how things work and how he fits into an organization. Once past those formalities, he is a delight, cheery and—for an astronaut at least—relaxed. But he likes to work, and so did his crew, Jack Lousma and Owen Garriott.

Lousma is as smiling and cheerful as Al, and even more gregarious. A football

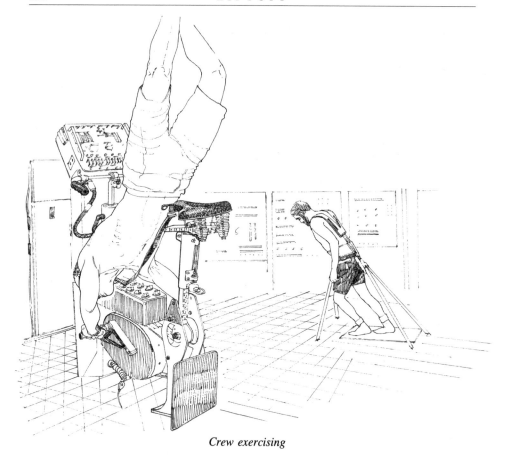

Crew exercising

player and aeronautical engineer from the University of Michigan, he was a major in the U.S. Marine Corps at the time. Jack is muscular and handsome, and projects an all-American image. Owen Garriott, the scientist of the three, is a wiry man with a serious look about him, but a pleasant sense of humor as well. To me he has an air of formality, and I am tempted to shake hands with him before we converse, even though we may have spoken only 5 minutes before.

The second crew tried very hard to pick up where the first left off, but they discovered, just as the first had, that adapting to weightlessness takes time, especially inside something as commodious as the workshop. Also, unlike Conrad's group, all three of the Bean crew suffered from motion sickness.

It was the first spaceflight for Lousma and Garriott, but Al Bean was the fourth man to walk on the moon, and during his Apollo flight with Pete Conrad he had felt no symptoms of sickness. Apparently the extra volume inside Skylab made the difference, allowing him a freedom of motion that agitated the fluid in his inner ear more severely than inside the Apollo Command Module. The Bean crew was also forced to spend parts of their early days troubleshooting and repairing balky equipment. But little by little they got into the swing of things, and by the fourth week they had not only adapted to space but were adept at besieging the ground with requests for more experiments.

After the flight they thought their 70-hour work week was insufficient to carry what they considered a proper load, and they suggested that more scientific experiments be added to the third flight.

Cheerful as they were, the second crew did find things to grouse about. Lousma was probably the best complainer of the bunch (he thought the wardroom table featured "the most miserable latch that's ever been designed in the history of mankind or before"), but he was also a genial host. He guided television viewers through such momentous occasions as the captain of the ship getting his hair cut.

"What's going on here? . . . Here's the leader of the mob, and here is the distinguished Professor Owen Garriott trimming his hair. Doing pretty well, too. He's flicking off the hair with that little blower up there . . . the tools of the trade used in Skylab, much like you might use on earth, a little comb with a razor on it. . . . We also have some bandages here, which will come out later when we have to patch him up. Well, I can see it's not going to be a professional job, but there's no waiting and the price is right. . . . Having your hair cut in zero gravity . . . you don't have to sweep the floor. The hair doesn't fall down, it doesn't even get on your shoulders. It gets up in the air sometimes. . . . That's what I'm doing, catching this loose hair with the vacuum cleaner. Doing a nice job, Owen. You might wonder why we chose Owen. . . . We figured you could always trust a barber with a mustache. . . . You know, there aren't many folks that get their hair cut at 18,000 miles per hour."

All the astronauts were impressed by the view from *outside* the workshop. When they went up front to the prow of the ship to retrieve film from the ATM, they had their best opportunity to get a wide-angle look at their home planet, a fundamental look at it without the intrusion of machinery into their field of view, except, of necessity, the edges of their helmet visors. They generally reacted with a feeling near reverence.

As Jack Lousma put it: "It's like being on the front end of a locomotive as it's going down the track. But there's no noise, no vibration; everything's silent and motionless. . . . To be on the end of the telescope mount, hanging by your feet as you plunge into darkness, when you can't see your hands in front of your face—you see nothing but flashing thunderstorms and stars—that's one of the minutes I'd like to recapture and remember forever."

Bean and Lousma were military pilots and tended to think alike, or at least Bean considered them a matched pair compared to the scientist Garriott. Owen saw things they didn't or at least saw them in a different light, and Bean respected his intellect and training. It was usually Garriott who made the suggestions that resulted in the crew working longer hours and conducting more scientific experiments. Owen himself turned the workshop into a classroom as he explained such fundamentals as magnetism. Holding a small bar magnet motionless in front of him, he explained what happened when he released it. "You see, it spins right over so that its axis is oriented with the earth's magnetic field.''

Owen was also adept at describing their environment in terms the other two might not use. Once the three of them gathered at the wardroom window to watch an aurora, a spectacular light show in the night sky. Bean and Lousma oohed and aahed, but the next day Garriott summarized it for Mission Control: ''The majority of it was closer to us than the earth's horizon. It was greeny in color. I couldn't see any red. The arc extended out away from the spacecraft toward the horizon, and the aurora tended to blend in with the airglow. Now above it, very faintly, you could see streamers, very thin striations in the aurora extending from much higher altitudes. Very thin rays, very dim, but thin rays more or less vertically aligned with the magnetic field could be seen.''

Toward the end Al Bean the overachiever suggested that the flight be extended, but because of medical conservatism and doubts as to the adequacy of the food remaining on board, Mission Control decided to save any extensions for the third mission. The controllers felt that they had gotten all they could expect from the three men.

As Flight Director Neil Hutchinson summarized it: ''We send up about 6 feet of instructions to the astronaut's teleprinter . . . every day . . . at least 42 separate sets of instructions, telling them where to point the solar telescope, which scientific instruments to use. . . . We lay out the whole day for them, and the astronauts normally follow it to a T. What we've done is we've learned how to maximize what you can get out of a man in one day.''

The third crew caused Hutchinson and his compatriots to eat these words. Sometimes they complained to Mission Control good-naturedly (''I understand you're going to teleprint up the Old Testament tonight''), but at other times their comments were biting and sarcastic as they poured their unhappiness into the B channel. All three were space rookies. Jerry Carr, the commander, was a lieutenant colonel in the Marine Corps, a mechanical engineer from the University of Southern California, and a fighter pilot. The crew scientist was Ed Gibson, a New York native with a Ph.D. in physics from the California Institute of Technology. Air Force Lieutenant Colonel Bill Pogue had a master's degree in mathematics from Oklahoma State, had flown with the Thunderbirds precision flying team, and was a graduate of the Royal Air Force's Empire Test Pilot School. In their official crew picture they are smiling even more radiantly than Conrad's or Bean's group, and that is how Neil Hutchinson considered them: an extrapolation of the first two crews, to stay up longer and to work more productively.

Their flight got off to a miserable start. Pogue got sick and vomited while still in the Command Module, and the three of them decided to conceal that fact from the ground. Unfortunately for them, they discussed the situation with a tape recorder running, and their conversation was telemetered down, where it was typed and distributed to all and sundry. The news was not taken lightly. It was exactly what the moguls of Mission Control disliked most: the crew taking independent action, not only ignoring the ground, but concealing information from them. "We won't mention the barf . . . we'll just throw it down the trash airlock." Sacrilege! Vomitus was to be freeze-dried and returned to earth for medical analysis. This crew had to be set straight, and quickly.

The chief of the astronaut office, Al Shepard, got on the radio and chewed a little: " . . . you made a fairly serious error in judgment here in the report of your condition." Jerry Carr didn't argue: "Okay, Al, I agree with you. It was a dumb decision."

All this, of course, was duly noted by the press, who had a fine time interviewing flight controllers about their loss of control over the crew. For a military man who was both a pilot and unit commander, to be told publicly that he had made a fairly serious error in judgment was a heavy blow to the solar plexus. For Carr and company, it began an adversarial relationship with the ground that continued for at least half their 84-day flight.

Although no one realized it, their worst enemy at the beginning was friendly, agreeable Al Bean, who had gradually built his crew into a model of weightless efficiency, zipping around the workshop 16 hours a day. Neil Hutchinson and his controllers seemed to have forgotten that it had taken Bean's crew about 30 days to reach this peak. Carr was expected to take over where Bean left off, and although some break-in period was allowed, the initial workload was clearly excessive.

The crew reacted like normal human beings: they started making mistakes, and the harder they tried, the worse things got. After their reprimand from Shepard, they squashed their initial tendency to complain, but after a week Carr did report on the B channel that "the best word I can think of to describe it is frantic." He went on to say that just learning to move around inside the workshop took a great deal of time. "I think you could tell by our voices that we were very, very frustrated. . . . No matter how hard we tried, and how tired we got, we just couldn't catch up with the flight plan. And it was a very, very demoralizing thing to have happen to us."

After 2 weeks Carr noted, "We got up here, and we let ourselves just get driven right into the ground." After 3 weeks he complained that his crew wouldn't "be expected to work a 16-hour day 85 days on the ground, so I really don't see why we should even try to do it up here." By the end of the fourth week Pogue told some experimenters that the problem lay with the schedulers: " . . . they're going to have to give us time to get from one point in the spacecraft to another. . . . I don't know how we're going to get that across . . . unless you . . . put your foot down and stomp it hard." By the fifth week it was Gibson's turn: "I personally have found the time since we've been up here to be nothing but a 33-day fire drill."

The flight controllers heard these complaints, but they really didn't want to back off very much. They always thought that the crew was on the verge of a breakthrough into new levels of productivity, and it was a matter of pride with them to squeeze as hard as they could. They, too, were being pressured, because this third Skylab flight was the last chance for an army of experimenters to get their precious data. New ideas and requests kept pouring in as time began running out. Exacerbating the problem was the fact that a "day off," scheduled one in ten, wasn't really a time for leisure, just a day of somewhat reduced activity.

Finally, in the sixth week, the crew decided to take a real day off, doing whatever they pleased. For Ed Gibson, it wasn't much different from a work day, for he spent the entire time at the ATM, peering at the sun. He was a solar physicist and had written a book about the sun. More than that, he was the first solar physicist to examine the sun without the distorting effects of the earth's atmosphere, and he was going to make the most of his rare opportunity.

The other two, the pilots, spent the day at the wardroom window, looking down on the earth as if from the cockpit of a high-flying airplane. They took pictures, but pictures of whatever appealed to them, not of locations on a list provided by Mission Control. Jerry Carr also taped a 6-minute message on the B channel, asking for a frank appraisal of the mission's status at the halfway point. This in turn resulted in an hour's conversation on the A channel, in which a real meeting of minds took place between spacecraft and ground for the first time in this flight. The ground, in looking at their data, discovered that the crew was now performing at the level of Bean's crew at the same point in their flight. The ground also agreed to provide more time for physical exercise, as the crew had requested, and promised to try to keep one hour quiet early in the morning and another one before bedtime.

These events and agreements seemed to clear the air, and the second 6 weeks of the flight proceeded more harmoniously and productively. At the 59-day point, two-thirds of the way through, they not only passed Bean's endurance record but they were right on schedule in regard to the experiments program. In the process, Mission Control learned that some people resent being treated as robots, or as cogs in their ground-based machine. As Pogue put it, "When I tried to operate like a machine I was a gross failure. . . . We've got to appreciate a human being for what he is." Treated properly, Jerry Carr, Ed Gibson, and Bill Pogue turned out to be as innovative and productive as their predecessors.

Still, they were ready to come home. When Mission Control inquired about an extension of their flight, they demurred, noting that they were running out of food. Actually, Bill Pogue told me years later, "we could have scraped up enough for another 2 weeks." It was just that they felt they hadn't been consulted earlier, that they had not been included sufficiently in the arguments for an extension, that they were being treated like the last cogs in the machine that needed oiling. "If they had started buttering us up 2 weeks ahead of time, if they had said, 'We have several experiments, and this is the last chance for 10 years. Just think about it,' " Pogue thinks they would have agreed.

At any rate, down they came on schedule after 84 days, the first splashdown that the television networks declined to show live, presumably because they feared low ratings. For the crew, returning to the earth's gravity was a real shock. Pogue, for example, unpacked a camera to photograph their parachutes, but found it so heavy he clutched it, unable to return it to its storage locker. It felt "like it weighed about 35 or 40 pounds."

The scientific experiments from Skylab can be divided into three categories: solar, earth resources, and human. The ATM provided new information about the sun that is both voluminous and arcane. To this layman the most impressive thing about the sun is the magnitude of the upheavals boiling up from within this ball of hydrogen gas that beams us the energy that permits and nurtures life on our planet. Ed Gibson describes a solar storm: ". . . I got my first look at the prominence speeding outward through the corona. . . . Fascinated, and with an overwhelming sense of awe, I watched this new feature grow. In less than an hour it was several hundred times as large as earth." *Several hundred times* as large as our entire planet! If an image of the earth is superimposed on the sun's disk, it looks like a flea on a basketball.

Yet according to astronomers, our sun is a very ordinary star. It's middle-aged (about 4.6 billion years old) and middle-sized, although it is classified as a dwarf star, in deference to giants such as Antares that are 500 times its size. It has no permanent features, the surface changing constantly as its gases bubble, swirl, and rotate. Temperatures deep inside the sun are high enough to trigger nuclear reactions, the most important being the process we have duplicated in the H-bomb, the fusion of hydrogen atoms to form helium. In the process, energy is released in the form of gamma rays. As these gamma rays make their way up toward the surface of the sun, they are slowed down and converted into infrared and visible light. This heat and light erupting from the surface begin an 8-minute voyage to earth. The earth, small as it is, intercepts only *one billionth* of the energy that spews out in all directions from the sun, but that is more than enough fuel to keep the earth machine operating. For example, the solar energy hitting just 1 square yard of earth is sufficient, if harnessed efficiently, to light and heat a small room. The sun is expected to continue this process for another hundred billion years.

The great value of the Skylab observatory was its ability to focus powerful telescopes on the sun *above the atmosphere* of earth and to make repetitive measurements across all pertinent wavelengths. Solar astrophysicists now have an accumulation of data (200,000 photographs!) that is still being assimilated. The sun's corona, that filmy outer layer, can only be photographed from earth during periods of total eclipse, but Skylab could get those pictures every day, and of the entire front disk of the corona, not just the edges protruding during eclipses. Furthermore, the Skylab cameras recorded the ultraviolet and X-ray bands, wavelengths crucial to understanding the sun's processes but blocked out by the earth's atmosphere.

Skylab flew during a period of solar passivity, but even the "quiet" sun featured a fascinating series of events: coronal holes and swirls, prominences, sunspots. Skylab recorded six large solar flares and over 100 minor ones. Skylab data documents in great detail the complex relationship between the separate layers of the sun's atmosphere, called the photosphere, the chromosphere, the transition region, and the corona. Perhaps more important, it offers new insight into the interaction between matter and magnetic fields, using data that shows the powerful magnetic forces in coronal holes.

Skylab also recorded events called coronal transients, perhaps the most spectacular of solar shows, wherein great clouds of electrons burst loose from the corona and stream out across the cosmos. These gargantuan bubbles have been blown past the earth for millennia, yet remained virtually unnoticed before Skylab. The impact of all this solar activity upon terrestrial climate, and such side effects as auroras, magnetic storms, and interrupted communications, continue to be studied. Skylab has changed the path of solar research. Scientists are eager to fly another ATM, or something like it, during the maximum activity phase of the sun's 11-year cycle.

As soon as humans reached orbit, they looked down in wonder. First it was simply the immensity, clarity, and beauty of the view of their home that fascinated them. But later, once the shock had worn off, they began to consider in a scientific sense the advantages of a high platform circling the earth each hour and a half. Drought, snowfall, ocean currents, earthquakes, volcanos, ice pack movements, pollution sources and distribution, crop management, prospecting for oil and minerals, tornado warnings, locating plankton—all these can be seen or helped by a satellite.

Humans need not be on board such an observatory, but their presence has proven to be valuable in a number of ways. An unmanned vehicle must be programmed in advance, and follows that program blindly, impervious to changing circumstances. Human crew members can discriminate, using their experience and judgment to adjust to their environment. If they see something interesting, they photograph it, regardless of their instructions. For example, the Carr crew saw and shot a volcanic eruption on one of the uninhabited Galapagos Islands a full day before the first ground-based photographers could reach this inaccessible site. They also repaired cameras, something no unmanned satellite can do.

Skylab was manned during summer, fall, and winter in the northern hemisphere. Conrad's crew concentrated on getting the workshop operating properly and did not photograph the earth in as systematic a fashion as did the others. Bean's crew surveyed 35 sites of geologic or ecological interest, and Carr's group took 2,000 photographs, accompanied by over 800 verbal descriptions. They found that they got better color discrimination when they photographed at high sun angles, and better definition of topographical relief at low sun angles, when shadows were more visible. Certain features, such as the San Andreas Fault, are so large that they are ideally suited to photography from orbit. Others, such as the development of tropical storms and the creation of ice fields, are not only large but also time-dependent, and the repetitive nature of Skylab's orbit enabled their maturation to be studied sequentially and in great detail.

Skylab's most useful photographs, from a scientific point of view, are of the oceans and the deserts. Oceanographers have long been frustrated by the immensity of their medium and the feeble tools they command. Of necessity, they take micromeasurements of a macrosubject. Dr. Gifford Ewing has observed that studying the ocean by lowering sensors into the water from ships is like reading the newspaper by putting it flat on the floor, sticking pins into it, and reconstructing words from the ink stains on the pins. Photographs from orbit at least allow the front page to be read.

Carr's crew was fascinated by the patterns they saw in the South Atlantic, where the relatively warm Brazil current and the cooler Falkland current flowed alongside each other for 2,000 miles, intertwining paths but never showing any mixture across their boundary. This area had been largely unstudied before, and the bright green color of the Falkland current as it flowed northeast from Cape Horn was totally unexpected.

The crew also spotted reddish plankton blooms in various locations around the world. The presence of plankton, those tiny animals that are the base of the oceanic food pyramid, is a reliable indicator that higher forms of sea life may be present. Both the Bean and Carr crews learned to use sunglint, the bright reflection of sunlight on water, to detect the ocean's surface texture and to spot eddies and areas of upwelling. These features on the front page are indicative of in-depth stories below. The ocean is in constant motion, and Skylab's ability to revisit at periodic intervals was immensely helpful. The Gulf Stream, for example, may shift its position significantly in 1 or 2 days. Other changes are slower, but 84 days was more than enough for the third crew to photograph ice in the Gulf of St. Lawrence throughout the winter of 1974, a period of rapid growth and dispersal of windblown ice floes. Such information can be of great value in routing ships through ice-infested waters.

On land the focus was on the world's deserts, especially on eolian (wind-related) sand deposits in the Sahel, that transition zone in northern Africa south of the Sahara and north of the Savanna of Senegal, Upper Volta, and Nigeria.

The Sahel (which means "boundary" or "shore" in Arabic) is an arid area where starvation is commonplace today, perhaps because of agricultural patterns that need to be changed. Certainly Skylab photographs alone cannot reverse this situation, but they did record dust storms, burning of foliage, and land use patterns that warn of aridification (the precursor to desert formation), loss of agriculture, and starvation.

The photographs also provide a better understanding of how sand deposits are affected by wind, rainfall, and natural barriers. The first step to real assistance for areas like the Sahel is a worldwide assessment of which deserts are natural and which are caused by man, and how. Or perhaps the first step is simply awareness of the problem. As Jerry Carr put it, "Most of the guys come back with an interest in ecology, for they see how much . . . desert there is, and how hard it is for the people who have to live there. . . . I think this mission . . . it's going to increase my awareness."

The third target of scientific investigation was man himself. What were the effects of spaceflight on his mind and body, and how did they change with time?

Skylab was the first American attempt to make fundamental research in biology and medicine an integral part of a space mission, and the durations of 28, 59, and 84 days were chosen primarily with the health of the astronauts in mind. Man has evolved in his present form in large part because of the pull of the earth on his body. Take gravity away, and what might happen?

If you want to know what you'd look like in space, lie down flat on your back, prop your feet up on a chair, and hold a mirror up in front of your face. You'll notice a definite change, an instant face-lift, as the loose flesh of your cheeks and neck—depending on your age, of course—sags less noticeably. Also, your eyes look different, more squinty, almost oriental. If you lie there for a few minutes you'll be aware of a fullness in your head, caused by extra blood that has drifted up from your legs. Your complexion may look a bit more florid.

If you were in space you'd notice these changes right away, plus you might develop reddened eyes and a stuffiness in the nose, like a mild allergy or cold. These changes are trivial and wouldn't affect your performance in any way. If you brought along a tape measure you'd discover that with no gravity to compress your vertebrae, your body had stretched a bit, and you might be as much as 2 inches taller than you were back on earth. (Pete Conrad liked this effect, because for the first time he was as tall as his wife Jane, but of course only as long as he stayed in space and she on earth, an unsatisfactory arrangement I presume.)

While these obvious changes are taking place in your body, there are also complicated processes beginning deep within you. Inside your thorax, sensors discover the presence of blood that gravity usually pulls down toward your legs. Your body now considers this blood "extra" and begins a complex process to get rid of it. It secretes hormones that have a diuretic effect, causing increased urine flow and body water loss, and a decrease in blood plasma volume. But although your body is now satisfied with slightly less blood flowing through it, it senses that the reduction has taken place by a loss of plasma, not red blood cells. It now considers the blood too rich in red cells and sets about reducing them. It does this not by destroying any cells but by curtailing the production of new ones.

Red blood cells are created in your bone marrow, but what exactly triggers the reduction in their output is not known. Another strange thing is that some of your red blood cells will begin to change their form. Normally they are disc-shaped, like doughnuts with a thin center instead of a hole, but now some of them look ragged around the edges, almost like a starfish. They are called echinocytes; why weightlessness produces so many of them, no one knows, but not to worry, they will disappear when you get back home.

Next you may become ill. Even if you've never been sea- or airsick, you may become aware of your stomach now, and it might be advisable to move slowly and avoid head movements. Chances are about fifty-fifty that you'll feel slightly nauseous for a couple of days, and don't be surprised if suddenly you have to vomit. Just carry a bag with you. In all likelihood you'll be able to continue with your work, and you can rest assured that you'll feel great in 3 days, or 4 at the most. You're in good company. Remember that Bill Pogue, who used to fly jet planes upside down with the Air Force Thunderbirds, and who was known as "old lead ear" because of his tolerance to preflight vestibular tests, felt like hell for the first few days aboard Skylab.

Medical experts still can't explain why some people get sick and others don't, but Joe Kerwin thinks the process is caused by a sensory conflict, that is, the brain receives different signals from the eye and the inner ear. "When you are below deck on the *Queen Elizabeth*," he says, "your inner ear detects motion, but that picture on the wall is steady. Ditto inside a spacecraft. When you leave your stateroom on the Queen, and go up on deck 'for a little air,' what you are really doing is looking at the horizon. That makes you feel better. The brain only takes about 3 days to figure this out."

As the weeks go by, you feel fine weightless, but your body has not yet adapted completely. Subtle changes are still taking place. Some you can control, but not others. The main thing you can do is prevent deterioration of your muscles by exercise. The heart muscle in particular needs help. It's no problem as long as you stay in space, but wait until gravity grabs you again.

Joe Kerwin, who only exercised for half an hour a day, was in worse shape after 28 days than Carr, Gibson, or Pogue after 84. They averaged 1½ hours a day, working on the bicycle ergometer and a treadmill. The basic problem is that on earth, gravity keeps you fairly fit, even if you do not exercise. Gravity forces your leg muscles to stay strong enough to support your upright body. Gravity causes your heart to labor as you walk upstairs. Even when you're sitting in a chair, gravity puts a load on your heart, as it pumps blood from your feet up to your head.

In weightlessness your legs have nothing to support, and their muscles atrophy. In similar fashion your heart gets lazy and actually shrinks in size. Work as hard as you can against mechanical resistance or on the ergometer and treadmill. Get that heart rate up and keep it there. You should be able to do just as much work up here as on the ground. If you want to know how you're doing, try the lower body negative pressure test. This involves a barrel-like device into which you insert your lower body and seal it tight with a rubber skirt around your waist. Then the air inside the barrel is vented to space, and the vacuum sucks the blood out of your head and torso and returns it to your legs, just as gravity would do on earth. The intensity and duration of the vacuum you can endure without fainting is a measure of your cardiovascular condition. Once Joe Kerwin "terminated the run early, based on a feeling of a little dizziness and cold sweat."

After 40 days in space your blood tests indicate that your bone marrow is once again producing red blood cells at a normal rate. Congratulations! But your urinalysis continues to be disturbing. It shows high concentrations of nitrogen, calcium, and hydroxyproline. The nitrogen means that despite your heavy exercise routine you are still losing muscle mass (you can see that your calves look skinny). The presence of calcium and the amino acid hydroxyproline indicates a more serious problem, the loss of bone. Your skeleton is less dense now, especially the bones of your lower back, legs, and feet. They will continue to lose 0.5% of their weight each month that you remain in space. Tests of bed-rest patients on earth suggest that after a year this bone damage may be irreversible. The calcium in the urine may also lead to kidney stones, a painful and dangerous ailment to have up here. Maybe it's time to call it quits.

When you return to earth, your first discovery will be something called orthostatic hypotension (low blood pressure due to body position). When you stand up, the blood will surge from your head and upper body down to your legs, a condition your body has not experienced while you were weightless. It will take a little while for your autonomic nervous system to relearn how to prevent pooling of blood in your legs. In the central veins of your legs there are flapper valves that close between heartbeats to prevent the blood from falling back down. These valves aren't working properly now. In addition, the muscles of your legs, which tighten and constrict the veins, are too relaxed and are allowing the veins to distend. Consequently, you will feel light-headed and perhaps need some help walking. Your leg muscles are weak, and your balance may be off a little. Your body senses that its blood volume is low, and you will be thirsty while it asks for replenishment. Your red blood cells will quickly return to their normal disc shape. Your tolerance for exercise will gradually return to preflight levels, and before long a medical examiner will be unable to detect any abnormality in your body due to spaceflight.

Your body will, however, be different in one respect. You have been exposed to protons from the Van Allen belts and from intergalactic cosmic rays. This radiation can have long-lasting, if poorly understood, aftereffects, although the medical community considers the dosage received by the Skylab crews to be well below the threshold of concern. NASA has established career radiation limits for astronauts. Jerry Carr, who received more radiation than the other eight on Skylab, could fly his mission once a year for 50 years before exceeding his career limit, which is based on a statistical analysis that accepts the possibility of life being shortened by 6 months to 3 years.

The nine men aboard Skylab were lucky in that the most serious physical ailment among them was Pete Conrad's dislocated finger, the result of his fooling around while exercising. In their 84 days, the third crew noted only some minor skin irritations and flatulence. ("Farting about 500 times a day," Bill Pogue noted acerbically, "is not a good way to go.") Only one of the nine, Joe Kerwin, was a medical doctor, and bandaging a dislocated finger was about as complicated as he could get with the equipment on board. I asked Joe what he would have done if Conrad or Weitz had had appendicitis. "Strap an ice bag on his belly, give him some antibiotics, and come on home."

Each crew member did receive about 80 hours of paramedical training, including instructions on how to perform a tracheotomy (cutting a hole in the windpipe) should some object stick in the throat and obstruct breathing. In weightlessness there are plenty of floating objects to be inhaled. There was a fancy dental kit, including pliers for extracting teeth, and 62 different types of medication. There was a microscope, and Joe Kerwin could analyze throat cultures. He could also do a basic urinalysis and such blood tests as red and white cell counts. Before the Skylab flights, various medical concerns were expressed, including the possibility of the celibate crew getting infected prostate glands that could lead to urinary tract problems. One doctor advised regular masturbation, advice Joe ignored.

In addition to the physical hazards of spaceflight, from the earliest days there have been concerns about what this bizarre new environment might do to someone's mind. Deep-sea divers have experienced a euphoria called the "breakaway phenomenon"; as they dive deeper, they become so entranced that they want to keep on going forever. Divers have been killed this way, although their deaths are probably attributable to the mixture of gases they were breathing (nitrogen narcosis) rather than to the visual splendor of their surroundings.

Nonetheless, psychologists worried about astronauts climbing ever higher and wanting to continue out into space, losing all emotional links to their home planet. I can attest that this is nonsense. The remote earth seen out the spacecraft window acts as a beacon, a magnet drawing your attention and your emotions. But Paul Weitz pointed out to me an interesting aspect of this attraction. It's to the entire earth that you are drawn, not to any one spot on it. It's not Houston, where your family is, or Washington or Moscow; you feel a kinship with, and a desire to return to, IT—the whole planet.

If nothing else, it was the wardroom porthole (Raymond Loewy's window) and the earth beyond that prevented feelings of isolation. It is true that the third crew did develop an "us versus them" attitude early in their flight, but this was because of their workload and the lack of frank communications with the ground, factors independent of the environment of space or the design of their workshop. Of all the nine, perhaps Bill Pogue felt stress more acutely than the others, and he was bothered by insomnia, which he attributes to stress. But again, it was probably stress due to being expected to put in a flawless 16-hour performance every day rather than stress caused by being in space.

Each crew got along well together. No special efforts were made ahead of time to assure psychological compatibility, nor do the Skylab crew members think such measures are necessary. Most, like Paul Weitz, resent any interference by psychologists or psychiatrists in what they regard as "operational" matters such as crew selection. I asked Paul about crew compatibility. "We've always pooh-poohed that," he replied, and went on to explain that if there were any real problems they would become obvious during preflight training, which can be quite stressful in itself.

Before a crew was formed, the commanders were asked if they could get along with their prospective teammates, and all the astronauts I have discussed this with agree that a commander should have veto power. They also indicate that the personality of the commander is extremely important. Weitz and Kerwin in particular are lavish in their praise of Conrad. "Pete was aces at building camaraderie" is the way Kerwin put it. "Our crew was loose, relaxed, talkative. Pete wouldn't let problems fester." Pete squawked at the ground occasionally, but when he did, he had constructive suggestions to make, and he was sympathetic to their problems as well as his own. Above all, Conrad is a communicator, and that seems to be the key to his success.

Several of the crew likened their time inside Skylab to a camping trip, and naturally enough their creature comforts assumed extra importance under these circumstances. Their sleeping and bathroom accommodations, the temperatures inside the workshop, food and drink—all these things weighed heavily on the mind. They generally gave Skylab high marks as a habitat, but not as a restaurant. Their menu was repetitive and became monotonous, despite the fact that their individual likes and dislikes had been considered. Jack Lousma: "I've asked myself every 6 days, whenever it turns up on the menu, 'How come I picked beef hash for breakfast?' " Owen Garriott, not a complainer, thought that the beef dishes tasted as if insecticide had been added to them. Everyone missed fresh food. But their most common complaint was that food in space tasted more bland than it did on earth. Caldwell Johnson thinks that the initial period of heat inside the workshop, before Conrad's crew rigged the sunshade, had something to do with this, exposing the food to an extra baking. Oddly enough, despite all their complaints the nine averaged the same caloric input aboard Skylab that they had established preflight. They ate as much but didn't enjoy it.

There were no alcoholic beverages on board. Conrad grumped that if he enjoyed a martini at home in the evening, why not do the same in space? After all, they were *living* there. At one time NASA was amenable to the idea of loading some wine on board. Apparently they thought whiskey too strong, and it had to be *fortified* wine so it wouldn't spoil. It would be rationed more parsimoniously than the British Navy's grog: 4 ounces per man each 6 days. Joe Kerwin had gone so far as to organize a fortified wine-tasting party at his home, and they had selected the winner by a blind taste test: Taylor's New York State sherry. Then, says Joe, "someone innocently mentioned it in a speech in the Deep South and letters began arriving at NASA." No booze on Skylab.

There was no sex on Skylab either. In talking to the Skylab astronauts I get the feeling that because of their heavy workload and their preoccupation with other matters, the absence of females was not a big problem for any of them. Bean likened it to being on a Navy ship. "No sense thinking about it," Weitz added. Kerwin thought about it less than on the ground. Bean considers the all-male arrangement superior in one respect: if no one is having sex, the resentment will be much less than with a mixed crew where not all are participating: "If some are doing it, you are going to want to. Hey—that fella's got a big smile on his face, and that bugs me." Carr thought about sex about the same amount inflight as on the ground, and Pogue thought it would be "a real problem" on a flight to Mars. Both thought that NASA could learn a lot from groups wintering in the Antarctic.

Developing Bean's point, they think mating pairs are fine, but only under certain circumstances. Jerry Carr says the Antarctic experience proves that the presence of women can be helpful, that it elevates standards of conduct and makes the whole thing easier to deal with. On the other hand, "pairing up can be a problem." If pairing takes place beforehand, or very early, it's accepted by the other group members, but "playing the field is bad news." Pogue echoes this point: "People will accept a sequestered couple, but there are problems with bed hopping, where there is a hope for sharing." I'd hate to be putting together a small crew—say five people to complicate it a bit—for a 2-year cruise to Mars and back.

Finally, were they fundamentally changed by their time aboard Skylab? No. "But it was a lot of fun" (Kerwin) and "a lot of hard work" (Pogue).

All three crews felt a twinge of sadness when they left Skylab, although all of them had had enough and were eager to be going back home. The moment was especially difficult for the third crew, not only because they had spent the most time there, but also because they were quite sure no one would ever again see what writer Henry S. F. Cooper, Jr., called their "house in space." When they departed, on February 8, 1974, they left it as one would leave a house, all tidied up and the pipes drained. They even left a package of food and film near the entrance hatch for future visitors. But there would be none.

The problem was that Skylab's orbit, 275 miles above the earth's surface, was too low to last forever. A scattering of molecules from the upper atmosphere, as well as particles coming from the sun, would bump into it, and little by little reduce its speed and start its descent toward the atmosphere—and a flaming death. Mission planners expected it to last 9 years, and they hoped the shuttle would be flying before then and could go up and nudge Skylab into a higher orbit. However, the shuttle got delayed, and Skylab's 9-year lifetime turned out to be 6. Ironically, one of the reasons NASA overestimated Skylab's life is that it didn't study the ATM data closely enough. Solar activity built up toward the end of the 11-year cycle and heated up the earth's atmosphere so that it expanded upward, putting more air molecules in Skylab's path. As a result, Skylab was scheduled to hit earth sometime in the summer of 1979.

As a functioning, productive laboratory, Skylab had elicited a great yawn from the public, but once it started to die, it made the front pages. It had successfully circled the earth nearly 35,000 times, but it was the final orbit that intrigued everyone. As if it were a roulette marker about to drop into a numbered slot as the wheel slowed, people bet when and where it would come down. Probably the ocean, but still . . . it did pass over all the great cities of the world.

Even Congress joined in the merriment. In the House of Representatives, Joint Resolution 356 was introduced by Jim Weaver of Oregon, authorizing President Carter "to proclaim July 1 through July 7, 1979, as 'National Sky(lab) Is Falling Week.' " "Whereas," it duly noted, "everything that goes up must come down; whereas, more than $2,600,000,000 was spent to put Skylab in orbit above the Earth . . . this . . . man-made meteor shower covering more than 400,000 square miles, consisting of more than 500 separate pieces, more than half of which will weigh in excess of 10 pounds, 10 pieces in excess of 1,000 pounds, and two pieces in excess of 4,000 pounds . . . should be properly observed. Furthermore, it is the sense of the Congress that Skylab should not be permitted to fall in Oregon, where it might mar the natural beauty of that most precious State.' '

It was Western Australia, not Oregon, that received the man-made meteor shower. There were reports of spectacular visual effects and loud noises as the pieces whizzed by in the predawn darkness, but no one was injured. A 17-year-old beer-truck driver from the village of Esperance claimed a $10,000 prize from a San Francisco newspaper for producing the first authenticated chunk of Skylab, some remnants of insulation from the workshop.

It was an ignominious end for that remarkable machine, our country's first space station, after a 6-year journey that covered nearly 1 billion miles. A few people still remember it. As Pete Conrad told me 16 years after he became the third person to walk on the moon, "If someone says 'space' to me, I don't think of the moon, I think of Skylab and all that volume inside the workshop."

The final flight of an Apollo Command Module took place in the summer of 1975, slightly a year after the last Skylab crew landed and 4 years before the workshop's demise. It carried three old friends of mine, and it conducted a rendezvous, not with Skylab, but with a Soviet Soyuz and its crew of two men, whom I would come to know later. In NASA jargon the flight was called ASTP, the Apollo-Soyuz Test Project. It was the result of President Nixon's and Premier Kosygin's desire to produce a tangible, technical, high-flying demonstration of the spirit of détente.

Before the two vehicles could dock, there were some engineering problems to work out, in addition to a host of political ones. Both the Apollo and the Soyuz crafts had docking probes, but they fit into drogues of different sizes and shapes. The two had different atmospheres, the Soviets using air at 14.7 psi and the Americans pure oxygen at 5 psi. Under these circumstances, even if the two could dock successfully, the hatch between them could not be opened

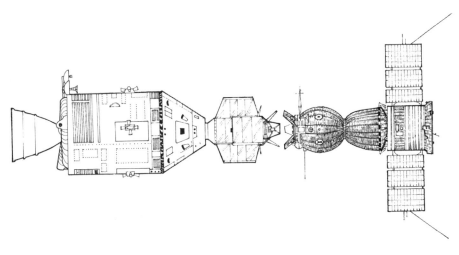

Apollo-Soyuz

because the Soviet crew would get the bends as the pressure in their cabin suddenly dropped, and the Apollo would be dangerously overpressurized. It took an interminable series of meetings in Moscow and Houston to decide how to solve these problems. First the Soviets agreed to drop their pressure to 10 psi. The U.S. agreed to construct an airlock module, and both sides agreed to build a universal docking mechanism, one that could also be used on future vehicles for rescue purposes.

The docking mechanism was an interesting design. Since neither side wanted to be relegated to—dare I say it?—the lesser role of the female, a probe (male) and drogue (female) arrangement was out. In its place, an androgynous docking apparatus was built, one that combined the "male" and "female" roles. Further, either the U.S. or the Soviet half could be the active partner, performing the docking maneuvers that locked the two together. It was like two hollow flowers embracing, each with three petals that clasped its mate. Although the two halves performed identical functions, the design habits of the two countries led the U.S. to use hydraulic shock absorbers, while the impact of the collision was attenuated on the Soviet side by electromechanical devices.

I liked the androgynous design a lot better than our lunar gear because, once docked, no probe had to be removed from the tunnel. The vital passageway remained clear. A more fundamental advantage was that future spacecraft could all be equipped with the same mechanism, and in theory crews could rescue one another by docking with a disabled craft.

Our people learned a lot about the Soviet system during the process of designing, building, and testing the docking mechanism. Our engineers found their Soviet counterparts to be friendly and competent, but it was like pulling teeth to get information from them. When asked, they would respond in the most superficial way they thought acceptable, and only as the Americans probed deeper would they reluctantly produce the required data. Furthermore, they seemed to be almost as reticent in sharing information with each other as with the Americans. Their internal lines of communication were sketchy at best, and the Americans soon learned that telling Ivan something didn't mean that he would spread the word to all who needed to know. Their organization was highly centralized, and even though their technical team leaders were easily capable of making decisions, they rarely did, preferring (or being required) to pass them on to higher authority. Once they received their marching orders, they showed a dogged persistence in getting things done. In short, the Soyuz team was a microcosm of the Soviet system.

The Apollo crew consisted of its commander, Tom Stafford, plus Vance Brand and Deke Slayton. Tom Stafford, who had been one of my instructors at the Test Pilot School, had one Apollo and two Gemini flights behind him, while Brand and Slayton were rookies. Vance Brand was a former Lockheed test pilot I used to arm wrestle with back in the early 1960s at the annual bash Lockheed threw for the Air Force test pilots at Edwards. Deke Slayton was one of the original Mercury astronauts, but after 16 years of NASA duty this was his first flight. The problem was that he had been grounded because of an erratic heart beat, and it had taken 10 years for the medics to decide that the problem was minor. At 48 he was the oldest man to have been picked for a spaceflight. For the 6 years I was an astronaut, Deke was my boss, and a fine one.

The Soyuz commander was Alexei Leonov. Whenever I hear the term "Russian Bear," I think of Alexei because he looks like a reddish-brown bear— compact, powerful, with a wrestler's arms. Once he greeted me with a world-class bear hug, lifting me off the ground and squeezing to the point that I started thinking about broken ribs. I'm not sure exactly what point he was making, but I think he was making one.

A man of many talents, Leonov was the world's first spacewalker. He is also a talented artist and a consummate PR man. His companion, Valeriy Kubasov, is quieter and more subdued than Alexei. Valeriy is short, rotund, pink-cheeked, easygoing, almost cherubic. I had about written him off as the Russian equivalent of a "good ole boy" when he presented me with a textbook he had written on celestial navigation. Even though it's in the Cyrillic alphabet, I can tell from the equations in it that it's far above my level. So much for "good ole boys." (I may have been conned, however, because Vance Brand tells me that it is standard Soviet procedure to have someone write an impressive document and stick a cosmonaut's name on it.)

The flight itself was picture perfect. The docking apparatus and airlock worked as advertised, and after what has been described as a heavenly handshake between commanders, the crews exchanged pleasantries and token gifts. Each spoke in the other's language, although it may have grated upon Russian ears to hear

Apollo and Soyuz

what Leonov good-naturedly referred to as Stafford's "Oklahomski." When information had to be imparted in a hurry, they quickly reverted to English, in which Leonov and Kubasov were quite fluent.

They also heard from their political leaders. A message from Leonid Brezhnev, General Secretary of the Communist Party, was read to them, including the statement "One can say that the Soyuz Apollo is a forerunner of future international orbital stations." Over a decade later, that still remains to be seen. President Gerald Ford spoke personally with each of the five men, asking questions about the docking and their food, and sending them his wishes for a "soft landing."

A product of détente, the cooperative spirit of Apollo Soyuz did not endure. As détente waned, so did the efforts made by each side in working toward new cooperative ventures in space. Despite the fact that the flight has not led to an international space station, most American participants look back on it fondly as a difficult technical achievement. They also believe that it was a worthwhile endeavor in that it proved the two countries *can* trust each other and work together with mutual respect—at least under some circumstances. And if we can breathe each other's air in space, why can't we _____? Fill in the blank.

Critics of ASTP say that the Soviets learned more from it than we did. It is true that the Soyuz was far less sophisticated than the Apollo CSM ("Soyuz was about halfway between a Mercury and a Gemini,'' says Slayton), but the Soviets could read all about Apollo in our unclassified literature. Stafford believes that ''the only thing they could have learned from us is management,'' but he doubts that they could apply the NASA style to their own endeavors. However, the Apollo crewmen do not belittle their counterparts or their space program. Vance Brand notes ''They are very determined people. If they decide to do something, they'll find a way. We shouldn't fear them, but we shouldn't take them lightly either.''

Today the crews make no special effort to stay in touch, except for the exchange of greeting cards at New Year's, but as Slayton says, ''At least we have some guys over there that we know and can talk to, and that's worth something.'' I think George Low, who was acting NASA administrator during some of this period, put it best: ''We live in a rather dangerous world. Anything that we can do to make it a little less dangerous is worth doing. I think that ASTP was one of those things.''

ASTP marked the end of both the Apollo and Saturn programs. With the exception of the deaths of Grissom, White, and Chaffee, the successes of these machines went far beyond anything I believed possible when I first began studying

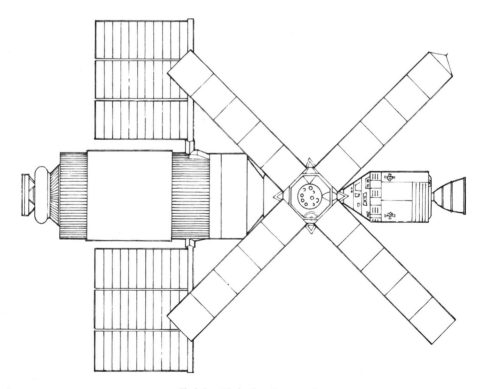

Skylab with both solar panels

them. To me the performance of Wernher von Braun's Saturn series was absolutely dazzling: 32 launches, including 15 manned flights, without a major failure. ASTP also was the end of a proud line of Mercury, Gemini, and Apollo spacecraft, and the end of an era of expendable manned launchers. As such, it was a major turning point, for the next U.S. space vehicle would be reusable, at least partially so.

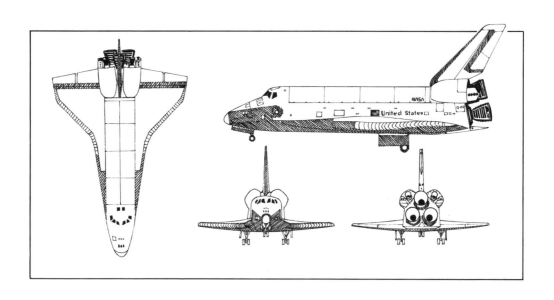

6

Wings and Wheels at Last

Skylab was tacked onto Apollo, more or less as an afterthought, but when Skylab was over, then what? NASA had been pondering its next major move long before the first lunar landing. As far back as 1967, the President's Science Advisory Committee had recommended that "Studies should be made of more economical ferrying systems, presumably involving partial or total recovery and use."

Two months after Apollo 11 returned from the moon, Vice President Spiro Agnew chaired a Space Task Group that offered a choice of three long-range objectives: (1) a full blown program costing $8-10 billion per year, including a manned Mars expedition, a space station orbiting the moon, and another, larger station in earth orbit that would be serviced by a shuttle vehicle; (2) an intermediate program costing less than $8 billion per year, but including a Mars mission; and (3) a program of $4-5.7 billion per year, for an earth-orbiting station and a vehicle to shuttle back and forth between the station and earth.

Agnew himself liked the Mars idea, and had tried to sell it in a nationally televised interview at Cape Canaveral just before the launch of Apollo 11, but Congress and the public turned a cold shoulder to the idea. The war in Vietnam was raging and there was no public support for a major national program such as a Mars landing. Consequently, in March 1970 President Nixon picked part of option No. 3, approving development of a shuttle vehicle but deferring the space station.

Thus the concept of reusability, technologically feasible and economically attractive, became the focus of NASA's future. Although disappointed that no king-sized program would replace Apollo, NASA engineers were intrigued by

Two-stage concept

the idea of flying one vehicle over and over. It implied a technological maturity, a feeling that space was here to stay, that launching ships into Kennedy's "new ocean" would become routine. And at least for an aeronautical engineer, the path to reusability was glorious.

A mature vehicle must not be plucked out of the ocean after every flight. That was not only expensive, but so awkward. Clearly, landing on a runway was in order, and to do that one needed the ability to maneuver, that is—WINGS! And the machine would come down on what? WHEELS! Wings and wheels, a dignified flying machine at last, no more awkward capsules with their puny lift to drag ratios, but a mature workhorse that would launch vertically, like its rocket predecessors, but fly back to a horizontal landing like a conventional airplane.

NASA spent 1970 and 1971 refining the shuttle concept. What emerged was a two-stage vehicle, totally reusable. The first stage, with a crew of two, would provide the initial thrust out of the atmosphere and then fly back to a runway, while the second stage continued to orbit. The second stage also carried two pilots, plus room for as many as 12 passengers on their way to a space station. The cargo bay would be big enough to accommodate space station modules. These rather grandiose plans, which would have cost over $10 billion to develop, were scrapped at the insistence of the Budget Bureau. Not being aeronautical engineers, the leaders of Congress and the Budget Bureau were not exactly overcome with joy at this new way of spending billions. The American public was preoccupied with the plight of the cities, monetary woes caused by budget deficits and inflation, and the revolt on many campuses, protesting the carnage in Vietnam.

In response to the demand for reduced costs, NASA redesigned the shuttle, which it started to call the STS, or Space Transportation System. The manned first stage of the STS was replaced by two gigantic solid rocket boosters (SRBs). The fuel for the second stage's ascent to orbit would be contained in an external tank (ET) attached to the belly of the manned vehicle, a delta-winged craft known simply as the orbiter. When empty the SRBs would be parachuted into the ocean and recovered for refurbishment and reuse. The ET would be jettisoned shortly before reaching orbit, would disintegrate during entry, and its remnants would fall into the ocean. The orbiter alone would continue into space, where it might remain for as long as 30 days.

These modifications to the original concept resulted in an STS that was not totally reusable, and one that had higher operating costs per flight, but initial development costs would be cut in half, to approximately $6 billion. Even so, NASA experienced great difficulty during the early 1970s in getting sufficient funds to pursue these plans aggressively. In 1971, for example, the Budget Bureau cut NASA's request in half, to $3.3 billion, and by 1974 the allocation had fallen to just over $3 billion.

The situation would have been worse except for some strong support on Capitol

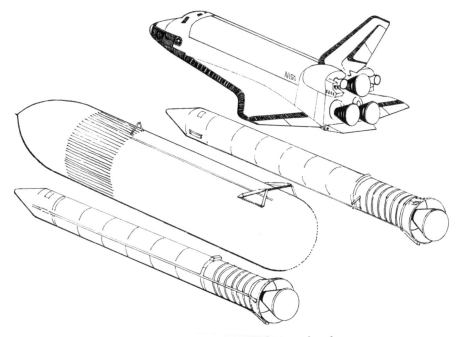

Space shuttle, solid rockets, and external tank

Hill. In the House of Representatives, the focal point for shuttle matters was the Manned Space Flight Subcommittee of the Committee on Science and Astronautics. The Subcommittee chairman was Olin Teague, a Democrat from College Station, Texas.

During World War II, "Tiger" Teague had landed at Normandy and fought his way across France and Germany as the commander of a Battalion of Infantry. In the process he gathered a chestful of medals for gallantry and was wounded three times, once seriously enough to keep him in the hospital for over a year. While still in the hospital, he decided to run for a vacant seat in Congress, and in August 1947, after a short campaign on crutches, he was elected. Tiger was a short, stout man, who limped badly because of his wounds, which had left one leg considerably shorter than the other.

When I knew him, during the period of confusion over the shuttle funding, he was still as combative as if a war had been declared over space funding. While a great supporter of NASA, he never hesitated to call the Agency's leaders to his subcommittee and "rake them over the coals," in the words of his successor, Congressman Don Fuqua. Teague understood that politically, Mars was an impossible goal, but when asked by Vice President Agnew what he did support, he replied "I know of one major contribution that can be made. That is the development of space vehicles that can be used repeatedly, with basic characteristics in common with transport aircraft."

Now NASA had an outspoken leader to champion its wings and wheels on Capitol Hill, and Tiger did it as he did everything else—forcefully. He publicly denounced President Nixon, " . . . he's allowed those damned pencil pushers in the Budget Bureau to set policy instead of following the experts' recommendations." He had to contend with fellow Congressmen like Ed Koch, who opposed space spending in general, arguing, "I just for the life of me can't see voting for monies to find out whether or not there is some microbe on Mars, when in fact I know there are rats in the Harlem apartments."

There was also opposition on his own committee. Congressman Joe Karth mistrusted the price tag attached to the President's Space Task Group Report: "NASA's projected cost estimates are asinine . . . NASA must consider the members of Congress a bunch of stupid idiots. Worse yet, they may believe their own estimates—and then we really are in bad shape."

Throughout the debate Tiger defended the shuttle, generally taking the long view when responding to his critics. "If Columbus had waited until Europe had no more internal problems, he would still be waiting, but the opening of the New World did more to revive European culture and economy than any internal actions could possibly have done." Although Teague was not pleased when NASA retreated to the pared-down version of the shuttle, using SRBs instead of a flying first stage, he nonetheless continued his pugnacious, and successful, stewardship during the critical years from 1973 through 1978.

In 1972, in an austere climate compared to Apollo days, NASA began awarding contracts for the STS. North American, now known as Rockwell International, was to build five orbiters, while Martin Marrietta won the ET contract, and Thiokol (later Morton Thiokol) the SRBs. Within NASA, managerial responsibility was divided among centers, with Houston's Johnson Space Center in charge of the orbiter, and the Marshall Space Flight Center at Huntsville, Alabama, responsible for the orbiter's three liquid-fueled main engines, the ET, and the SRBs. The Kennedy Space Center at Cape Canaveral would assemble the components, check them out, and conduct the initial launches. The Air Force would later prepare a launch pad at Vandenberg Air Force Base, California, for polar orbits.

In addition to its other responsibilities, Houston was designated the lead center for STS matters, a new concept for NASA, and one that didn't work out too well. For example, the SRB manager in Huntsville really had two bosses, his own Center Director plus the STS manager in Houston. There has never been much love lost between Huntsville and Houston.

Clearly the orbiter would be a flying machine of unparalleled complexity, combining in a vehicle the size of a small airliner the advanced technology enabling it to operate in the vacuum of space and land—without engine power—on a 10,000-foot runway. In the process it would pass through three distinctly different flight regimes: hypersonic, supersonic, subsonic—each with its own aerodynamic problems.

A most fundamental question was how to generate the lift required to maneuver back into the atmosphere, glide to a runway, and slow down enough to land at a safe speed. The shuttle was to be a practical "operational" machine that could fly routinely in and out of various orbits, and some of its proposed mission profiles required a lot of lift.

For example, consider a launch into polar orbit from Vandenburg, California, on a military mission. The Air Force wanted the orbiter to be able to take off, place a satellite in orbit, and land at Vandenberg after one turn around the earth. The shuttle would launch toward Baja California, pass nearly over the South Pole, continue around the other side of the earth, circle over the North Pole, and come back toward Vandenberg from the direction of Alaska. But guess

Polar orbit

what? During the 90 minutes required to complete one orbit, the earth would have turned 22½° on its axis, toward the east, so that Vandenburg would no longer be directly underneath, but would have rotated some 1,500 miles out from under the orbiter's flight path. To land at Vandenberg, instead of in the Pacific Ocean, the shuttle would have to bank left during entry and generate enough lift to glide all the way over to where Kansas would have been if the earth hadn't rotated at all. That's a lot of lift, and most of it at hypersonic speeds.

The Air Force also stipulated that the shuttle must be able to carry into orbit an object 60 feet long, 15 feet in diameter, and weighing 65,000 pounds. If for some reason this behemoth could not be released in space, then the orbiter would have to return *with* it. Here again lift would be all-important, to allow the heavily loaded orbiter to slow down enough for a safe landing on a runway. Some quick calculations indicated that the orbiter would require a subsonic lift to drag ratio of at least 4:1, relatively small for an airplane, but over 10 times that of the Apollo Command Module.

In the years since Max Faget's zero lift capsule had won out over Al Eggers' "half baked potato," a lot of work had been done on the "potato"; it had evolved into a series of "lifting bodies," small, one-man experimental craft that flew at Edwards during the 1960s and early 1970s. These bathtub-shaped machines proved that lift could be produced without wings.

NASA, the Air Force, and Boeing also worked on a delta-winged spacecraft called the X-20 Dyna-Soar. Although the Dyna-Soar never flew, it was the progenitor of the shuttle, for analysis indicated that a huge cargo bay was practically impossible to fit into a reasonably proportioned lifting body. Furthermore, the

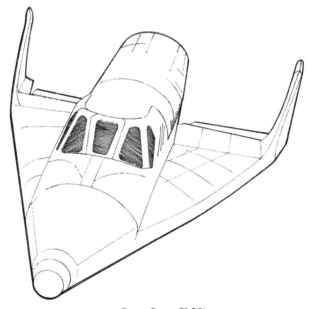

Dyna-Soar (X-20)

delta-wing design produced a higher supersonic L/D, better handling qualities, and the extra margins of safety in approach and landing. In this case, the wings were well worth their extra weight. Even with wings, however, the orbiter's final descent would scare the socks off the average glider pilot. It comes screaming down at over 300 mph, at an angle of 20°. At 1,500 feet above the ground, the pilot pulls back on the stick, causing a 1.7 G deceleration, and intercepts a shallow, 1½° glide path. At the last moment the pilot drops the gear and the orbiter touches down at over 200 mph. If the pilot makes a mistake and undershoots or overshoots, that's tough, he crashes, because at this point the orbiter is without engine power and cannot make a second try.

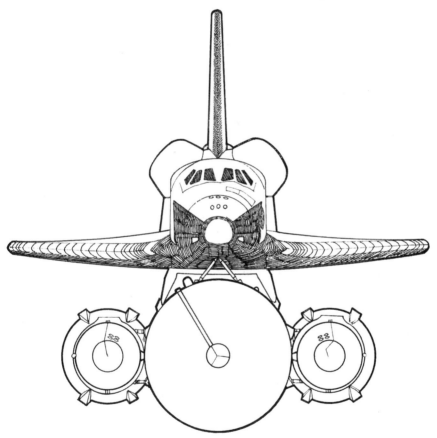

Bird's eye view of the shuttle

Landing on a runway was not the only challenge that faced the shuttle's designers. If a bird flew over a Mercury, Gemini, or Apollo on the launch pad and looked straight down, it would see a symmetrical shape, basically a bullet designed to pierce the atmosphere cleanly. A bird's-eye view of the shuttle shows a welter of shapes: first a big bullet (the ET) flanked by two smaller ones (the

SRBs). Tucked next to the three bullets are the stubby wings and flat belly of the orbiter, and on one side its almost square fuselage with an elongated dorsal fin jutting out. From an aerodynamicist's point of view, this configuration is a mess, and as it picks up speed during ascent, complicated things start happening.

In less than a minute after liftoff, the shuttle reaches "max-Q" and a force of around 700 pounds of wind blast is exerted on every square foot of the craft's cross section. A few seconds later, approaching Mach 1, shock waves form first on the orbiter's wings, and then on the nose of the orbiter, the ET, and the SRBs. These waves can bounce back and forth in the nooks and crannies between the various shapes, and cause extra loads on the structure that are difficult to analyze. For example, the shuttle is protected from entry heating by some 30,000 ceramic tiles glued to its skin. Composed of silica fibers, the tiles are lightweight but fragile. In the stressful zone around max-Q and Mach 1, it was feared a shock wave might blast loose a long row of tiles, as one would undo a zipper. Then during entry the aluminum structure beneath the missing tiles would melt.

There are other hazards during entry. At very high speeds in the upper atmosphere, the orbiter is not aerodynamically stable but depends on its computers for control. If they fail to keep its nose pointed straight ahead and if a sideslip angle of just a couple of degrees develops, the orbiter can enter a deadly hypersonic spiral from which the pilots cannot recover before the craft has exceeded either its structural or temperature limits. While the tiles do not take the pounding during entry that they do after launch, those on the orbiter's belly are exposed to temperatures of nearly 2,000°F.

But from a pilot's point of view, the trickiest part of the entry is judging the speed and altitude required to reach, but not overshoot, the runway. *Speed* is a measure of kinetic energy, *altitude* of potential energy. Speed and altitude taken together define the vehicle's total energy, and this is what must be rationed carefully in relation to the distance to the runway. The trick is to keep a little extra—but not too much—energy up your sleeve, and get rid of the surplus by using the speed brakes or making S-turns.

Because wind conditions are subject to last-minute changes, the people in Mission Control and the pilots must work together second by second to ensure that the shuttle dissipates just the right amount of energy as it flies precisely down its preplanned, three-dimensional path, dropping from an altitude of over 100 miles and a speed of nearly 18,000 mph over Australia to zero altitude and zero speed at the far end of the runway at Edwards, California.

In appearance the orbiter seems more an airplane than a spacecraft, in deference to the aerodynamic problems of entry and landing. The exaggerated height of its vertical fin and, especially, the complex sweep of its delta wing suggest a maturity of design, and in fact scale models of the orbiter have logged a record 50,000 hours in wind tunnels. Its fuselage is of necessity bulky, containing the 60 × 15-foot payload bay, reached through huge clamshell doors along its spine.

Upon closer examination, however, some of the features that allow this hybrid to operate in space become apparent. For example, reaction control nozzles are strategically located on the nose and tail to allow the orbiter to change its attitude in the vacuum of space, where elevons (a combination of elevator and ailerons) and rudder are useless. Its sheath of protective tiles is also a telltale sign that this is no ordinary airplane.

At the rear of the orbiter are five rocket motors, two small and three large. Both of the small ones, mounted in a pod on either side of the vertical stabilizer, produce 6,000 pounds of thrust. They are used to provide the final burst of speed to reach orbit, to change orbits, and to deorbit. They are quite conventional, burning hydrazine and nitrogen tetroxide, and are similar in design to the Gemini and Apollo thrusters, although larger. The three liquid fueled main engines are hardly conventional. Unlike their predecessors, they are designed to be reused—up to 50 times. They can also be throttled back to 65% or up to 109% of rated thrust. Yet, while they must be durable, they are souped up in that they operate at a chamber pressure of 3,000 psi, over three times that of the Saturn V's engines. Their turbopumps, about the size of large truck engines, each generate 70,000 horsepower. Their pumps spin at 35,000 rpm, three times faster than a race car's engine, and supply enough propellant (1,000 gallons per second) to empty an olympic size swimming pool in less than 2 minutes.

Not only do the three engines operate at very high temperatures and pressures, but they are of necessity clustered together in the tail of the shuttle, where a failure in one may very well spread to the other two. The disintegration of a high-speed pump, for example, might easily cause a lethal spray of turbine blade fragments and other metallic debris. At 370,000 pounds of thrust each, the orbiter's main engines are considerably smaller than the 1.5 million-pound thrust F-1 engines of the Saturn V, but they are still huge by most standards, larger than a Mercury capsule, and yet also lightweight (6,700 pounds each) and hence fragile. In my mind, the cluster of these three space shuttle main engines (SSMEs), is the Achilles heel of the STS design. Combustion is just too close to explosion.

The orbiter itself does not carry fuel for the SSMEs, but simply routes liquid hydrogen and oxygen to them through 17-inch diameter lines coming from the ET. The ET is the part of the reusable STS that is *not* reusable. Twenty-seven feet in diameter, it looks like a gigantic thermos bottle and it functions like one, its insulated tanks storing over 1.5 million pounds of hydrogen and oxygen at cryogenic temperatures. It is covered with 10,000 square feet of thermal protective foam. The ET is not flimsy, however, because on the launch pad it has the weight of the orbiter hanging from its side. The ET, in turn, is suspended between the two SRBs. During launch, it must act as a backbone and absorb the acceleration of the five engines. When the ET runs dry, shortly before reaching orbit, it is jettisoned and disintegrates as it enters the atmosphere. On a launch from Cape Canaveral its fragments come down in the Indian Ocean.

The SRBs, which flank the ET, burn from just before liftoff through the first 2 minutes and 8 seconds of flight and then are parachuted into the ocean for reuse. Producing 2.6 million pounds of thrust, each SRB is as tall as a 15-story building and contains over a million pounds of a rubbery solid propellant the consistency of a pencil eraser, a mixture of ammonium perchlorate, aluminum powder, iron oxide, and a binding polymer. These ingredients are mixed in 600-gallon bowls using the equivalent of a giant eggbeater. For ease of fabrication, handling, and transportation, the propellant is poured from these bowls at the Thiokol factory in Utah into four segments that are shipped individually by rail and assembled at the Cape.

Solid fuel rocket motors are simpler than liquid fuel rocket motors because a solid propellant is already in the combustion chamber, and does not need fancy machinery to pump it there. On the other hand, a combustion chamber for solid fuel consists of the entire casing, and in the case of the shuttle the combustion chamber walls were not made of one piece of steel, but of four segments pinned together and sealed with O-rings, a fatally flawed design.

Three other weaknesses in the STS design are more obvious than segmented combustion chambers or fragile SSMEs.

The first is that, unlike previous manned systems, the crew is not perched up on top of all the explosives; instead, the orbiter is mounted belly to belly with the SRBs and ET. There is a good reason for abandoning the traditional tandem configuration. If the orbiter were on top of the stack, its SSMEs would be used only as second stage engines, that is, they could not be fired until the SRBs below had been jettisoned. Not only would the SRBs have to be bigger in this case, providing all of the first stage thrust, but there would be no way to check out the SSMEs on the pad, to make sure they were functioning properly before committing to a launch. The orbiter was moved down to near ground level for reasons of efficiency, so that the thrust of its SSMEs could be used to augment the SRBs, and for reasons of safety, to allow the SSME's power to be checked prior to liftoff. But at the same time this decision put the crew compartment down where it would be near the center of an all-out explosion.

The second departure from traditional design involves the abandoning of an abort system during the early phases of flight. The orbiter can readily accommodate two ejection seats, and they were used for the first four test flights, but as the crew size grew, ejection seats became increasingly difficult to provide. It's not just the seats themselves, it's designing systems and hatches to allow 8 or 10 seats to catapult simultaneously through the sides of the crew compartment. An escape rocket that could pull the entire orbiter out of a fireball would be so big that lifting it off the pad would become a problem. If the orbiter were designed so that the crew compartment could be separated from the rear end of the shuttle and rocketed free, the escape rocket would be smaller, but such a system would still be very complex and would decrease the payload delivered to orbit by 15,000 or 20,000 pounds. In addition, escape pods in aircraft have not been very successful.

The orbiter is designed to be a mature, or operational, system and—just like in an airliner—for better or worse crew and passengers land in the same vehicle in which they take off. But, unlike an airliner, at low altitude the orbiter is attached to SRBs that, once ignited, continue to burn until their fuel is depleted, or they explode, or they are destroyed by ground command. Even if the SRBs could be turned off somehow, this early after take-off the orbiter doesn't yet have enough altitude to return to a runway; instead, it glides to a crash landing in the ocean. It is doubtful whether anyone could survive a ditching in the orbiter: it flies too fast and hits the water too hard.

Third is the problem of dead-stick landings. One of the original shuttle concepts included jet engines mounted on the inner surfaces of the payload bay doors. The doors could then be opened and the engines used for a powered approach to a runway, with a wave off if necessary. For reasons of weight and complexity, the engines were removed. Intensive training of astronauts and ground controllers has so far resulted in flawless energy management during descent, but ask any fighter pilot whether he'd like to end each flight with a dead-stick landing . . . no, wait a minute, fighter pilots are crazy enough to enjoy the idea. And most astronauts are former fighter pilots.

I tried it myself one time, in the shuttle landing simulator in Houston. Sitting in the commander's seat, I felt more like I was in a transport or bomber than a fighter plane or a spacecraft. The familiar Apollo Command Module type switches reminded me of the shuttle's heritage, but the commodious seats and large windows made it definitely feel like flying a big airplane. The scene out the window of the simulator, produced by a television camera in an adjacent room filming a huge painting of Edwards, California, was fuzzy, but quite realistic. I could see the runway easily, but I had some difficulty hitting the right spot on it. I got a definite sensation of speed as the scenery passed the windows, and of motion as the entire simulator rocked back and forth on gigantic pistons, in response to my control stick movements.

My problem was that I tended to overcorrect. Diving at 300mph, I had trouble maintaining my alignment with the imaginary runway as I began my pull out. My instructor laughed. "Sensitive, isn't it? You want to know something funny? Experienced pilots usually have more trouble with it the first time than beginners. They try too hard to control it, and don't let it fly itself enough." After a couple of runs I got the hang of it, but I still think the basic concept puts a lot of pressure on the commanders—not that most of them aren't used to pressure from their test-flying careers.

Behind the pilot and copilot and their main instrument panel are additional clusters of instruments and controls, and seats for two more crew members. In the rear of the flight deck are two windows facing aft, overlooking the payload bay. It is here that one of the crew uses special controls to maneuver a 50-foot-long manipulator arm that grapples payloads in and out of the cargo bay. Like a human arm, the manipulator has joints at the shoulder, elbow, and wrist. Below the cockpit, or flight deck, is the middeck, with accommodation for at least four additional crew members. Here also are the sleeping quarters, the food preparation center, the bathroom, and an assortment of equipment that varies from flight to flight, depending on the experiments to be performed.

An airlock leads from the middeck to the payload bay, through which astronauts pass for their extravehicular forays. On some flights a European laboratory module called Spacelab is carried inside the payload bay, and it is reached through a tunnel connected to the same airlock. On the ground the crew enters and exits the orbiter through a hatch in the side of the middeck. This hatch also contains a 10-inch porthole, the only viewing port in the middeck. But in the flight deck up above there are 10 windows; six pointed ahead, two overhead, and two to the rear. For the first time astronauts get a really panoramic view from orbit.

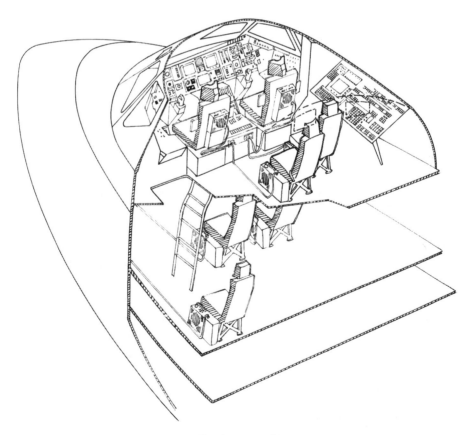

Shuttle crew cabin

The orbiter is also different from its predecessors in that its atmosphere consists of 79% nitrogen and 21% oxygen at a pressure of 14.7 psi, the same as at sea level. This mixture not only provides better fire protection than pure oxygen at 5 psi, but it also removes one variable from the equation when considering the effects of spaceflight, and scientists like it for that reason. If changes occur during a shuttle flight, they cannot be attributed to a different atmosphere. A nitrogen/oxygen mixture does, however, add one complication. The spacesuits, for ease of movement, continue to operate at the reduced pressure of 4 psi, so prebreathing of 100% oxygen is required prior to extravehicular activities in order to prevent the bends. The airlock is used to make the transition from 14.7 psi to zero without having to depressurize the cabin.

The systems aboard the orbiter are quite similar to those employed on earlier spacecraft or on airplanes. Three fuel cells the same size as Apollo's produce six times as much electricity—12 kilowatts at peak load. Hydraulic power, generated by three auxiliary power units, is used for airplane-like functions: elevons and rudder, the speed brake, landing gear, nosewheel steering, and wheel brakes. The speed brake is an unusual design, integrated with the rudder, which is made up of two identical halves. Hinged at their forward edge, the two rudders can open to form a drag-producing wedge that slows the orbiter as required during approach for a landing.

In a strange way the speed brake can be thought of as an engine: for example, if a "normal" approach is made using the speed brake one-half open, then the pilot has some flexibility in adjusting his glide path. If he wants to steepen it, he can open the speed brake fully. If he thinks he is going to land short, he can close the speed brake to shallow his approach, just as he would do by adding engine power, *if* he had an engine.

There are several indices of the shuttle's complexity. One is the number of switches, dials, and controls required to operate it—over 2,000 of them, far more than in any other flying machine. Second, the crew must be assisted by computers—five of them, with a total of 16 million bytes of information, capable of nearly 2 million operations per second. I always thought that the Apollo Command Module was about as much as I could handle. I really had the wind taken out of my sails by several of my astronaut friends, who in response to my questions generally guessed that the shuttle was the equivalent in complexity of four or five Command Modules.

Just consider the orbiter's flight controls. It has elevons and a rudder for flight through the atmosphere, and a reaction control system for changing attitude above the atmosphere. As the orbiter begins its descent, only the reaction jets are effective; near the runway, only the aerodynamic controls. In between, in the tricky hypersonic regime, the reaction controls must be smoothly phased out while the aerodynamic controls gradually take over. The computers perform this delicate balancing act, blending controls with a speed and precision that the most gifted pilot could not match. The orbiter is truly a hybrid machine, half airplane and half spacecraft, that must handle properly both in and above the atmosphere.

Colonel Joe Engle probably has had more experience flying high performance aircraft than any other astronaut. At Edwards, Joe flew the X-15 rocket research aircraft at speeds up to Mach 5 and altitudes above 50 miles. By virtue of these brief zooms into space, Joe was the only person to come to NASA already wearing astronaut wings on his military uniform.

Given the size of the shuttle, says Joe, it flies remarkably well. "It feels like a good, solid flying machine, a big airplane that tries to be responsive to the controls. It's an electric airplane." By electric Joe means that the pilot's hand is not linked directly to elevons and rudder, but that his hand controller sends electrical impulses that are converted into mechanical forces. Pilots also call this "fly-by-wire." Fly-by-wire airplanes are the wave of the future, but they tend to feel a bit artificial to pilots accustomed to the sensation of direct feedback from air flowing over the control surfaces.

The first shuttle test flights were not orbital, nor even suborbital, but investigations of the last 2 minutes of flight. They were performed by the Enterprise, a vehicle fully sized but without engines and unable to fly into space.

In 1977 Enterprise was launched five times from a piggyback position atop a Boeing 747 over Edwards, gliding down from 25,000 feet to a landing first on the dry lake bed and then on a runway. I asked Joe Engle about these tests. Years of study, months of training, for less than 2 minutes of flight? Ah, that was the challenge, Joe said, to cram as many maneuvers as possible into the few seconds available, to get the most valuable information. "The reward came from developing the flight profile, to arrange things for maximum efficiency."

The landing itself? Joe points out that the huge elevons on the rear of the shuttle can make landing tricky. For example, if the pilot jerks back on the stick just above the runway, that action can have a reverse effect, that is, it causes the elevons to raise and momentarily dump some lift, causing the craft to drop slightly and land hard. Compared to flying a conventional aircraft, the pilot must plan his approach meticulously, and avoid making large corrections at the last moment.

Returning from orbit adds an extra ingredient. After a week or so in space, the pilot finds he is extrasensitive to G forces when he returns to earth. In the landing pattern, when he pulls 1 ¾ Gs turning into final approach, it "feels like 6," says Joe. During the landing itself, the pilot must rely almost entirely on his eyes, and disregard his usual seat-of-the-pants sensations, which have temporarily become oversensitized by prolonged weightlessness. Each shuttle pilot practices at least 500 landings in training aircraft before trying the real thing.

Extensive ground check-out followed the Enterprise's brief flights, including vibration tests performed on the entire mated assembly of orbiter, ET, and SRBs at Huntsville, and SSME firings on Launch Pad 39A at the Kennedy Space Center.

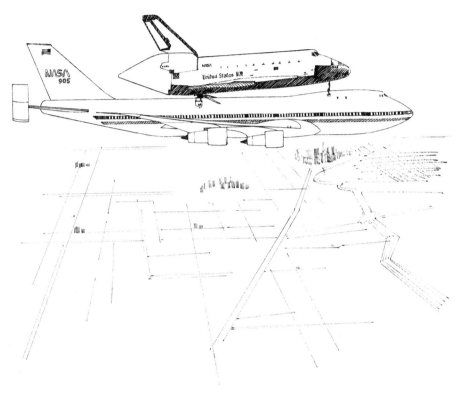

Space shuttle-747

These SSME tests were crucial, because these engines had failed with some regularity during earlier tests, including such potential disasters as fuel lines that cracked, combustion chambers that burned through, and turbopumps that disintegrated. In 1978 a National Academy of Science inquiry concluded that NASA had underestimated the engine breakthroughs required, and that budget pressures had unduly abbreviated the component testing process. Components had not been adequately tested as separate entities before the engines were assembled and tried as a unit.

The 30,000 ceramic tiles glued to the orbiter's skin were another source of concern. They were cut precisely to various shapes and thicknesses depending on their location, and attached by gluing them to the orbiter. It was a very time-consuming process, and ended up taking over 300 man years of labor (figured at 40 hours a week) on the orbiter Columbia. This gigantic number is due in part to the fact that many of the tiles had to be removed and reglued, and then ''pull-tested'' individually to make sure they were secure enough to withstand the rigors of max-Q and entry heating.

By early 1981 the STS was ready to fly to orbit, more than 2 years behind its original schedule and about $1 billion over budget. Columbia was first up,

John Young

to be followed by Challenger, Discovery, and Atlantis. The commander of STS-1 was my old buddy John Young, now Chief Astronaut after two Gemini and two Apollo flights, and first in line to try the new flying machine. The orbital flight test program was carefully crafted to include more than 1,000 different tests and data collection procedures.

The satisfactory performance of some components was verified simply by their normal operation, while others—such as the manipulator arm—were put through a rigorous series of drills. On STS-1 John Young and Bob Crippen stayed up for 2 days and verified the basic ability of the orbiter to perform a simple mission. Their flight was such a success that, after landing on the Edwards lake bed, the normally reticent Young performed a little dance around the orbiter, flapping his arms and giving the thumbs-up signal. It was a celebration, he told me, of the fact that none of the tiles had come unglued from Columbia's belly.

STS-2, 3, and 4 continued the test sequence, putting the electrical, guidance, navigation, communications, and environmental systems through their paces. The flight equipment worked very well, and the major modification to emerge from the flight tests was the injection, on the pad, of water into the rocket exhaust plumes to reduce acoustic shock to the lower parts of the vehicle. The STS-3 flight was the longest of the series, staying up 8 days, and it showed some versatility by landing at White Sands, New Mexico, when the lake bed at Edwards became flooded.

At the completion of the fourth flight of Columbia the STS was declared "operational," in the sense that payload requirements would henceforth take precedence over spacecraft testing. The use of the term operational, however, is susceptible to other definitions, and as late as February 1984, a National Security Decision Directive declared that the shuttle would not become "fully operational" until 1988, or when it could launch 24 missions a year. NASA's early predictions were that there would be 116 flights between 1981 and 1985, instead of the 23 actually flown. Although the STS was the only U.S. launcher available for large commercial payloads, the commercial market did not expand as rapidly as anticipated, and much of it was lost to the French Ariane rocket, a formidable competitor.

At any rate, the operational, or partially operational, phase of the STS program began with the flight of STS-5 in November 1982. It was the first to carry a crew of four and the first to fly without ejection seats. By the 10th flight, planners were bold enough to abandon Edwards and land Vance Brand's crew on the specially constructed 3-mile-long runway at Cape Canaveral. Although the runway at the Cape is as long as the one at Edwards, there is no dry lake bed to provide a safety margin, in case the wind shifts suddenly or the brakes fail. Early mornings at Edwards are usually clear and calm, while at the Cape a fog bank off the coast may roll in at the last moment. Like the weather, the condition of the runway itself can change capriciously, as deer, alligators, and wild pigs stroll back and forth across it. These ground critters are only a minor annoyance, Vance Brand told me, because they can be shooed away, but "those big buzzards drive around between 500 and 2,000 feet," and they can truly be a hazard.

Vance's mind, however, was not on the fauna but on the aerodynamics of his approach. Despite approximately 3,000 practice landings in training aircraft and one actual shuttle landing at Edwards, he felt a bit strange this time. "When I got over central Florida, between St. Petersburg and Orlando, I wondered how I was going to get down in time. I had a feeling of great height and speed. Then I went into that spiral staircase." Vance likens a shuttle approach to descending a steep spiral staircase. With a lift to drag ratio of only 4:1, the orbiter glides like a stone during its final turn, and from 90,000 feet he was on the runway within 6 minutes. When he coasted to a stop, he tried to get up out of his seat and found he couldn't. After 8 days in space, he needed about half an hour to "shake out my legs" before climbing down the Challenger's ladder to terra firma.

Of the four orbiters, Challenger has flown most frequently, logging nine successful round trips. Columbia, which made the first five flights, later underwent

Landing at the Cape

modifications and has only recently rejoined the fleet. It has made seven trips. Discovery, making its debut in August 1984, has flown six times, and Atlantis has two flights to its credit. Prior to January 1986, the STS flew 24 successful missions in 5 years, an average of one flight every 2½ months. It has carried as many as eight people and stayed up as long as 10 days.

Ninety-six men and women have flown aboard the shuttle: 88 Americans, three West Germans, and one each from Canada, France, Mexico, The Netherlands, and Saudi Arabia. Seventeen men have flown as commander, 25 men as pilot. The remainder of the 96 astronauts, including eight women, have been mission or payload specialists, the terms given to those who specialize in the on-board equipment, payloads, and experiments. In addition to aviators and engineers, crew members have represented a wide variety of scientific disciplines, including physics, geology, astronomy, and medicine. Many people have made more than one flight; Bob Crippen has made four. There have been 15 spacewalkers, including one woman, Kathryn Sullivan. Senator Jake Garn of Utah and Congressman Bill Nelson of Florida have also ridden the orbiter.

Joe Allen, a former Yale University physicist turned astronaut, has written about the orbital phase of flight: "In spite of its airplane-like appearance, being aboard an orbiter as it circles the earth is in no way like being aboard a hypersonic airplane . . . it's more like being aboard a ship. . . . There is no airplane noise, no steady drone of jet engines, no constant vibration of the fuselage, floors, and seats. Instead there is the reassuring and gentle hum of pumps, fans, electronics, and the occasional click of a valve and quiet hiss of nitrogen or oxygen being fed into the cabin to replenish the gases that have escaped. . . . The direction that the orbiter actually points is controlled by the crew using an autopilot, coupled to gyroscopes and a set of 44 rocket thrusters located in the nose and tail of the spaceship. Six of these thrusters are verniers, tiny rockets of only 25 pounds thrust, small enough to be held in one's hand. The verniers . . . although small, hold the 200,000-pound ship silently and precisely in a fixed pointing direction, an invisible anchor in the sea of space. In sharp contrast . . . are the 38 primary thrusters, large rockets with 870 pounds of thrust. . . . When a primary thruster fires, a tongue of flame shoots out several feet . . . with a sound like a cannon going off. . . . The orbiter shakes noticeably with each salvo. . . . Performing a series of maneuvers using the primary thrusters must surely be like being aboard a fighting ship in the last century."

Although the STS has not flown as often as NASA hoped, it has performed a wide variety of tasks and experiments. Several dozen satellites have been carried aloft and released. Most of them have been communications satellites on their way to geosynchronous orbits, far higher than the 300-mile altitude the orbiter can sustain. Nonetheless, the STS has performed effectively as a first-stage booster for satellites that carry their own propulsion stages. In 1985, for example, the shuttle placed in orbit 11 communications satellites for eight organizations from five countries. The few failures experienced by these satellites have been due to upper-stage propulsion malfunctions, or other causes having nothing to do with their initial placement by the orbiter.

The STS has also begun to show promise as a satellite retriever. In February 1984, for example, two $75 million communications satellites were deployed by Challenger but failed to reach their desired orbits, due to identical failures of their upper-stage motors. Nine months later Discovery conducted a rendezvous with each of them, and two extravehicular astronauts muscled them into the payload bay. They were then returned to earth for refurbishment and future use. Satellites designed for easy retrieval, which these were not, could make such operations fairly routine in the future. Also the planned development of an unmanned, remote-controlled utility vehicle will give the STS a greater capability to reach high-flying satellites and return them to the orbiter.

In 1984 Challenger not only retrieved a satellite but repaired it *in situ* and returned it to useful service. Solar Max, valued at $77 million when it was launched in 1980, had failed in orbit. Two spacewalkers spent 4 hours in the payload bay working on Solar Max, replacing a failed control system and a faulty electronic box, and then relaunched it. Today it continues to transmit information on solar flares, at a cost of $48 million in repair bills instead of an estimated $235 million replacement cost. Some future satellites are now being designed with their vulnerable components positioned for easy replacement. The more expensive the satellite, the more sense it makes to keep it operating by in-orbit maintenance.

Four of the most promising shuttle flights carried a laboratory in the payload bay. Called Spacelab, and financed by the European Space Agency through

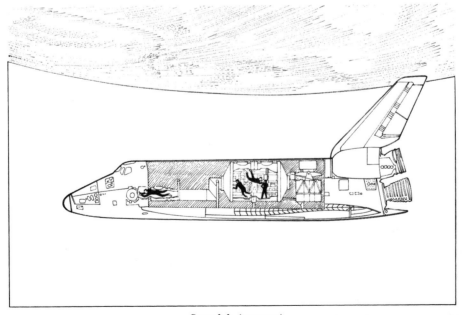

Spacelab (cutaway)

a cooperative agreement with NASA, this habitat has allowed scientists and engineers to investigate how to use space either to turn a profit or to better our knowledge of ourselves and our universe. Commercial possibilities seem brightest in the fields of materials processing, which includes purifying drugs, making larger and more perfect crystals for the electronics industry, and manufacturing new alloys in weightlessness. Atlantis has carried samples of a hormone, called erythropoietin, that can be used to treat red blood cell deficiencies such as anemia. Purified in weightless conditions by a process called electrophoresis, this product will next be used in animal tests and could have widespread human applications.

The orbiter has served as a workshop for a host of other activities. Astronauts using maneuvering units have ventured more than 300 feet away from it. Earth measurements have been taken in the fields of hydrology, agronomy, and oceanography. An animal colony has even been flown, giving life scientists an opportunity to examine organs and tissues exposed to weightlessness.

There is also a military dimension to the shuttle. This country's military and civilian space programs have always overlapped to some extent. The Atlas and Titan boosters, for example, were not designed to launch Mercury or Gemini spacecraft, but intercontinental ballistic missiles for the Air Force. Military experiments were flown aboard Gemini spacecraft. The early astronauts were nearly all military pilots on active duty, and several hundred officers were loaned to NASA from the Defense Department during the Apollo program. Some, such as General Sam Phillips, served in key management positions.

When the decision was made to build the shuttle, it was designated *the* Space Transportation System, the one and only way to put heavy payloads into orbit, and the message to the military was clear: climb aboard the shuttle. The Air Force balked, arguing that putting all its space eggs in the shuttle basket was imprudent and indeed irresponsible, and the White House eventually backed off to the extent that the Air Force was allowed to purchase a few Titan III boosters as insurance against a grounding of the shuttle fleet. When the Air Force needs a spy satellite, it doesn't want to wait for NASA to get around to launching it.

In 1985 two shuttle missions dedicated exclusively to the Defense Department were flown. I watched one of them go, from Titusville (12 miles west of Cape Canaveral), as close as the public was allowed to the classified launch. There was a news blackout, causing much press unhappiness, but the crowd around me seemed undisturbed by the fact that they were not allowed closer or that they didn't know when liftoff would be. As part of the blackout, the radio announced only that launch would take place between 1:15 P.M. and 4:15 P.M.. Conversation focused on this puzzle, and a number of bets were made. Finally the radio gave us a break and announced that launch would be in 7 minutes. We squinted into the pure blue sky and watched the tiny launch pad on the horizon. In the last few seconds, we started a people's countdown, ragged but cheerful. The first sign that Launch Pad 39 had heard us was a twin plume of light brown smoke shooting out on either side of the shuttle. It was the three SSMEs that ignite seconds prior to liftoff. Then a brilliant flash of light, almost pure white at its source—the solids—expanding into a widening almond-yellow cone at liftoff.

Its ascent was more rapid than the Saturn V but still slow enough to register clearly in eye and brain. The shuttle's exhaust was almost too bright to watch. It was also almost too beautiful to watch. As the wind distorted it, the thick white trailing plume became a vertical cloud, with swirls and eddies on its edges, a curved scimitar slicing into the sky with the incandescent shuttle on its tip. At the instant the SRBs cut off, 2 minutes and 8 seconds after launch, the plume ceased and only a dot of light, a miniature sun, remained. It faded rapidly to the east, and the crowd began to disperse.

The shuttle doesn't quite have the ground-rumbling, gut-shaking power of the Saturn V, but it's close. Watching it on television is no substitute. You really have to be there, to see its unfiltered fire and listen to its crackling roar.

Compared to the limitations imposed by earlier throw-away hardware, the reusability of the shuttle has resulted in a versatility of space operations, just as its proponents claimed it would. Its cost, on the other hand, has stubbornly resisted NASA's best efforts at reduction.

In the early seventies statements by the President and the NASA Administrator indicated that the STS would bring about a dramatic decrease in launch costs. Comparisons between then and now are difficult, because of inflation and changes in the structure of the program, but roughly the cost per pound delivered to orbit, in 1970 dollars, was estimated to be $150 then and $450 now. Or, in today's dollars, it costs approximately $1,400 per pound to deliver an object to low earth orbit.

This calculation assumes a nearly full payload bay and a launch price of $80 million, which is about what NASA has been charging its commercial customers. NASA's critics say that the Agency should charge at least $160 million, to recoup research and development costs fully and to assume a fair share of NASA's overhead burden. If that were done, the price per pound to orbit would be about six times NASA's original estimate.

One large miscalculation was the fact that STS operations have been a lot more labor-intensive than planners envisioned. Originally, an orbiter was expected to be able to fly at 2-week intervals, but the quickest turn-around time demonstrated to date has been 2 months. After its first flight, for instance, Columbia was found to have over 400 nicks, scrapes, and gouges in its fragile tiles. Most flights end at Edwards, and the orbiter has to be transported back to Cape Canaveral on the back of NASA's sole 747, a time-consuming process and one that has itself resulted in further tile damage. Spare parts have not been adequately funded, and as a consequence time has been lost cannibalizing one orbiter to provide parts for another.

A contributing factor to the high cost per shuttle launch is that the market for commercial payloads has not developed nearly as rapidly as NASA had forecast. During the seventies NASA predicted that by the mid-eighties a shuttle would be launched approximately weekly. Instead, the busiest year to date, 1985, saw only nine shuttle flights. Unfortunately, salaries and other overhead expenses don't decrease in proportion to the flight rate. Furthermore, the French Ariane rocket has proved to be a tough competitor, with the French government subsidizing it to the extent that the charge to launch a communications satellite is only $30 million.

The STS story described so far is what might be called Phase I; an awkward and tardy beginning, with some good ideas but not enough cash to develop them quickly or fully, finally achieving the flexibility of a reusable vehicle, but still at too dear a price, and with some compromises of safety. Yet Phase I was by most standards a success story, with two dozen useful and productive flights between 1981 and 1983. If Phase I of the STS were tacked on to Mercury, Gemini, Apollo, Skylab, and ASTP, the record would show that over a 25-year period NASA had successfully flown 133 men and women in space without an inflight loss of life.

All that changed abruptly on January 28, 1986. I was sitting in my office that morning, without a TV set, the shuttle far from my mind, when I got a call from my wife Patricia. "The Challenger just blew up," she told me, "you could see the whole thing."

The flight lasted 1 minute and 13 seconds. Liftoff was normal, but less than a second later cameras recorded a puff of gray smoke coming out of the side of the right Solid Rocket Booster (SRB). Three seconds later the smoke disappeared and all appeared well until 58 seconds after liftoff, in the region of max-Q, when a flickering flame appeared from the same part of the right SRB.

The flame quickly grew into a large plume that impinged on the surface of the External Tank (ET). At 65 seconds a change in flame color indicated that hydrogen from the ET was mixing with the fire, which continued to grow until 72 seconds.

At that point, at an altitude of 48,000 feet, a series of events occurred extremely rapidly: the lower strut connecting the SRB to the ET broke loose, and the aft portion of the hydrogen tank separated. The SRB hit the forward portion of the ET, rupturing the oxygen tank. Massive amounts of hydrogen and oxygen burned, enveloping the orbiter in a fireball. The Challenger's reaction control fuel ignited, and the characteristic reddish brown smoke of nitrogen tetroxide could be seen on the edges of the main fireball.

The orbiter broke into several large sections, one of them the forward fuselage containing the crew of seven. It tumbled out of control and hit the surface of the Atlantic Ocean 2 minutes and 45 seconds later, at a speed of slightly over 200 mph. It is not known whether the crew were dead or alive when their compartment hit the water, but that 200 G impact was far in excess of anything the human body can withstand.

Challenger explosion

Neither the crew nor the flight controllers were aware that anything was wrong until the explosion, although there were some subtle clues in the information telemetered to the ground. Sixty seconds after liftoff, for example, the chamber pressure of the right SRB was noticeably lower than its twin, intimating that there was a leak in it somewhere. Conversation between Challenger and the ground was normal in all respects. The only indication of any concern on board was one brief exclamation on the intercom by Mike Smith, the pilot. "Uh-oh," he said at the same instant that the fireball enveloped them. Radio contact was lost at this point.

My first reaction to my wife's phone call was, "It's finally happened," although I didn't tell her that, because around my family I've always pretended that bad things like that are not going to happen. But if someone had suggested to me in 1963, when I first became an astronaut, that for the next 23 years none of us would get killed riding a rocket, I would have said that person was a hopeless optimist, and naive beyond words.

In a machine of such immense power, such immense *lightweight* power, a thin and fragile barrier separates combustion from explosion. It was not the solids that I mistrusted so much as the cluster of three main liquid fuel engines, their turbopumps churning away at 35,000 rpm, their thin-walled combustion chambers burning vast quantities of high energy hydrogen at 6,000°F, hydrogen that seconds before had been stored as a liquid at minus 423°F. But beyond any single component was simply the complexity of the total machine. Surely such a behemoth could not be expected to work flawlessly time after time, year after year. Surely we humans have not become infallible just because we assemble, test, and fly these machines.

The public reaction to Challenger was much more evident than it had been 19 years earlier when Grissom, White, and Chaffee were burned to death. One reason, I suppose, was that, as my wife said, "You could see the whole thing." And see it we did—across the television screen in living color, again and again, backwards and forwards. A second reason was the composition of the crew.

The commander, Dick Scobee, and the pilot, Mike Smith, were traditional astronaut types, white male test pilots. But the other five were different and perhaps in a way that drew the public to this crew, almost as if the people next door were on the flight. Ellison Onizuka, one of the three Mission Specialists, was an Air Force flight test engineer, and also the first American of Asian descent to fly in space. Ron McNair was an MIT physicist and the second black American in space. Judy Reznik, with a doctorate in electrical engineering, was the second American woman in space. The last two, called Payload Specialists, were not NASA employees, but were on board for reasons not connected with operating the shuttle. An engineer and satellite designer, Gregory Jarvis, was to conduct experiments concerning how fluids in space reacted to various motions. The seventh crew member was Christa McAuliffe, a high school social studies teacher from Concord, New Hampshire, who had been selected as the first teacher-in-space from a list of over 11,000 applicants.

The seven were a microcosm of American society, and watching their spacecraft being blown to bits was like witnessing a tiny, but vital, piece of this country being destroyed.

Within a week of the accident, President Reagan appointed a commission to investigate its causes and recommend corrective action. "The members shall be drawn from among distinguished leaders of the government, and the scientific, technical, and management communities." The Chairman was William P. Rogers, a Washington lawyer and former Attorney General and Secretary of State. The Vice Chairman was Neil Armstrong, and the members included Brigadier General Chuck Yeager, the first person to break the sound barrier, Dr. Sally Ride, first American woman in space, and Dr. Richard Feynman, a Nobel prize-winning physicist.

After the Apollo fire in 1967, the investigation was left in NASA's hands, but this time the White House decided that an outside look at NASA was necessary. Although several NASA support teams were organized to assist, of the Commission members only Sally Ride was a NASA employee. It was a good group, and one that worked very hard. Their report, produced in the relatively short period of 4 months, is a clear and comprehensive analysis of the technical and managerial aspects of the accident.

The Rogers Commission's job was made easier by the fact that, from the very beginning, all available evidence pointed toward the aft field joint in the right SRB as being the sole cause of the accident. The Commission considered other possibilities, such as leaks from the ET or failure within one of the three main engines, but could find no indications of these from the wreckage recovered from the ocean floor. On the other hand, pieces of the right SRB corroborated the fact that a failure had occurred in the joint between the two lower segments—the aft field joint. Portions of the two O-ring seals that are designed to prevent hot gases from leaking through the joint were destroyed, allowing a fiery jet to escape and to impinge on the adjacent surface of the ET.

To understand the complex sequence of events that contributed to this failure, it is necessary to examine the design of the joint.

An SRB arrives at the Cape in four segments that are assembled there (the "field"). The joint between the bottom and next-to-bottom segments is called the aft field joint. The bottom segment is positioned vertically and its mate is lowered onto it. The two are joined by the insertion of a protrusion around the rim of the upper segment, called a tang, into a deep groove in the lower segment, called a clevis. Once tang and clevis are in place, pins are inserted to hold them together. The inside of each segment's steel wall is lined with insulation, which separates the casing from the solid propellant itself. Prior to joining two segments, a zinc chromate putty is applied to the insulation. Joint sealing is provided by two flexible, rubber O-rings, each ¼ inch in diameter.

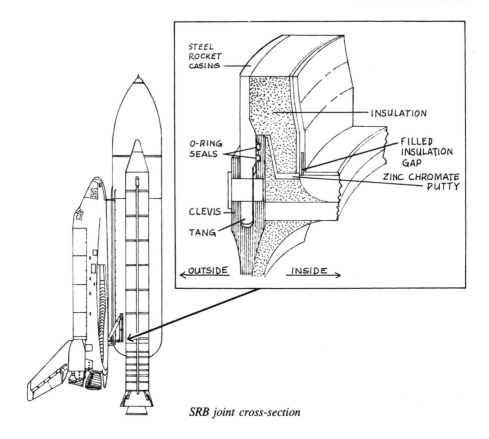

SRB joint cross-section

The idea is that when the propellant ignites, the rapid rise in pressure within the casing causes the putty to be squeezed outward toward the O-rings. Trapped air in front of the putty forces the primary O-ring into the gap between tang and clevis, and seals the joint. The putty also acts as a thermal barrier, shielding the O-rings from direct contact with the combustion gases. The secondary O-ring is a backup, performing a function identical to the primary ring if for some reason hot gases slip by the putty and the primary.

According to the Rogers report, there are a number of factors that influence the joint's ability to effect a safe pressure seal. A segment may be damaged or out of round. The gap between tang and clevis may be too small or too large. An O-ring may lose its resiliency at low temperatures. Ice may form in the joint. The putty's performance can affect the timing of the sealing process, or cause O-ring erosion.

To complicate matters further, at liftoff there is an interaction between the thrust of the three liquid engines and the two solids. Six seconds before liftoff, the liquid engines are started and brought up to full power. Over a million pounds of thrust tries to lift the orbiter off the ground but cannot do so because the solids are still bolted solidly to the pad. Consequently the thrust bends the whole stack—orbiter, ET, and SRBs. The astronauts feel this on board as a lurch forward (they call it a "twang") and the joints in the SRBs feel it as a bending force, with compression on one side of the booster and tension on the other.

Then one quarter of a second before liftoff the solids ignite (with over 5 million pounds of thrust) and balance is restored as the shuttle begins its climb. But the point is that the O-rings must seal not just under static, laboratory conditions, but during the dynamic bending process that produces rapid shifts of load at launch. During ascent, additional side loads are caused by shifting winds, especially in the vicinity of max-Q.

To make a proper seal, the O-ring must be squeezed quickly into the gap between tang and clevis because that gap widens as combustion gas pressure rises and deflects the joint. It takes only half a second for this gap opening to occur. One would think, therefore, that the smaller the initial gap, the better. However, a smaller than normal gap can also have a harmful effect by causing the O-ring to get caught between tang and clevis and squeezed out of shape before ignition. Then when needed to fill a rapidly expanding gap, the flattened O-ring may instead allow gases to pass by it, and not spring back into shape fast enough to be of much value as a seal.

Temperature is very much a factor in O-ring resiliency. The air temperature at Challenger's launch was 35°F, but 4 hours before, the Rogers Commission calculated, the temperature at the aft field joint had been only 28°F. At this temperature an O-ring has lost most of its ability to spring back to a round cross section after being flattened. Another factor to consider is that Challenger experienced 7 inches of rain during the 38 days it sat on the launch pad. Almost surely some water found its way into the U-shaped clevis. Ground tests have proven that ice frozen downstream of the secondary O-ring can prevent it from seating.

The two SRB segments involved in the Challenger accident had been used before and their diameters had grown slightly because of this wear. Calculations indicate that at the time of launch the average gap between their tang and clevis was .004 inches. Also the segments were not exactly round, so that in some places the gap was less than .004 inches, resulting in the primary O-ring being compressed to a considerable extent. At ignition, the gap increased rapidly due to combustion pressure and bending loads, creating a gap that the frigid, flattened O-ring was not able to fill. Under these conditions putty alone cannot contain the pressure, and a leak path connecting the rocket's combustion chamber with the primary O-ring quickly developed. The functioning of the second O-ring was impaired, either by ice or by a process similar to the failure of the primary ring, and the hot gases then had a path through which to begin penetrating the side of the SRB.

A series of puffs of smoke was observed coming from the aft field joint of Challenger's right SRB between 0.7 and 2.5 seconds after SRB ignition. The puffs appeared at a frequency of three per second, an interval that matches the cyclic opening and closing of the tang-to-clevis gap due to the dynamics of engine firing. The puffs then disappeared until 58 seconds after liftoff, when a flame reappeared in the same area. It is possible, says the Rogers Commission report, that the leak was continuous but unobservable during this interval, or that the O-ring gap might have been resealed by combustion debris until 58 seconds, when wind loads caused some disturbances to the vehicle. In any event, concludes the Commission, " . . . the cause of the Challenger accident was the failure of the pressure seal in the aft field joint of the right Solid Rocket Motor. The failure was due to a faulty design unacceptably sensitive to a number of factors. These factors were the effects of temperature, physical dimensions, the character of materials, the effects of reusability, processing, and the reaction of the joint to dynamic loading.''

The Challenger's fatal flight was preceded by 24 successful shuttle missions. Since the SRBs were recovered from all these flights, NASA and the SRB contractor, Morton Thiokol, had 48 post-flight opportunities to disassemble and examine field joints. The results of these inspections indicated that in about a dozen cases hot gases were reaching the primary O-rings and blowing by them. As far back as the second shuttle flight, in 1981, a primary O-ring was found to be partially eroded. The worst case was found on a 1985 flight, with evidence that both SRBs had experienced extensive primary O-ring blow-by. Black soot and grease were found between the primary and secondary O-rings and not just in one tiny spot, but in an 80° arc around one and a 110° arc around the other. Also, for the first time, a secondary O-ring showed the effect of heat. This flight was launched with a joint temperature of 53°F, the coldest prior to the Challenger accident.

NASA and Thiokol did not exactly ignore this evidence of a marginal design, but neither did they fix the problem (except for the minor action of changing to a new putty in the summer of 1983). What they did was mostly on paper: study the problem, organize a task force, write memos. A near-term suggestion was to increase the size of the O-rings, but there were objections, and it was never done.

In August 1985, Thiokol submitted to NASA a list of 43 possible design concepts for field joints, but again nothing was done, despite the fact that concern was mounting within Thiokol. "Help!" began one memo concerning the pace of the task force's work. As each shuttle flight approached, the problem was discussed, but the presence of the unblemished (except in one case!) second O-ring swayed the argument in favor of allowing one more launch. After all, had not the field joints held together each time?

As Commissioner Feynman put it, the decision making was "a kind of Russian roulette . . . the shuttle flies . . . and nothing happens. Then it is suggested, therefore, that the risk is no longer so high for the next flights. We can lower our standards a little bit because we got away with it last time. . . . ''

On the afternoon of January 27, 1986, a mid-level manager at NASA Huntsville asked his Thiokol counterpart whether Thiokol felt any concern about the

Challenger being launched the next morning in what might be subfreezing weather. A 1-hour meeting of Thiokol engineers at their Utah plant expressed real concern, and their fears were relayed to Thiokol's man at Cape Canaveral, who informed NASA. As the afternoon progressed into evening, the matter escalated within both the NASA and Thiokol hierarchy, and culminated in a teleconference between a large group headed by four Thiokol vice presidents and their equivalents from NASA Huntsville, several of whom were physically located at the Cape in anticipation of the launch.

Charts were displayed that traced the history of O-ring erosion and blow-by. The data showed that the O-rings performed more sluggishly as the temperature decreased, and that the worse case seen so far had been the coldest—at 53°F. Temperature at launch the next morning was predicted to be 26°F.

Robert K. Lund, Thiokol Vice President of Engineering, recommended that the Challenger launch be delayed until the temperature reached 53°F. George B. Hardy, Deputy Director of Science and Engineering at NASA Huntsville is reported by Thiokol personnel to have said he was "appalled" by this recommendation, but that he would not recommend launching if Thiokol did not. Lawrence B. Mulloy, Manager of SRBs at Huntsville, has been quoted as saying, "My God, Thiokol, when do you want me to launch, next April?" although he says this statement has been taken out of context. At any rate, for whatever reasons, Thiokol Vice President Joe C. Kilminster asked for "5 minutes" to discuss the matter privately within the Thiokol group.

The 5 minutes stretched into half an hour as the group again went over the effects of temperature, with the lower level engineers unanimous in their opinion that the launch should be delayed. The senior man present, Senior Vice President Jerald Mason, "turned to Bob Lund and asked him to take off his engineering hat and put on his management hat," according to one participant, Roger Boisjoly. "I felt personally that management was under a lot of pressure to launch, and that they made a very tough decision, but I didn't agree with it. . . . The determination was to launch, and it was up to us to prove beyond a shadow of a doubt that it was not safe to do so. This is in total reverse to what the position usually is in a preflight conversation of a flight readiness review," Mr. Boisjoly concluded.

After the accident, Commission Chairman William P. Rogers asked Robert Lund: "How do you explain the fact that you seemed to change your mind when you changed your hat?" Mr. Lund responded, "We had to prove to them that we weren't ready, and so we got ourselves in the thought process that we were trying to find some way to prove to them that it wouldn't work, and we were unable to do that."

At 11:00 P.M., the teleconference resumed and the Thiokol management in Utah stated that they had reassessed their position, that the temperature effects were a concern, but that the data were admittedly inconclusive. Joe Kilminster read a launch rationale, and George Hardy requested it in writing. It was sent by telefax, and said that although colder O-rings were harder and took longer to seat, "if the primary seal does not seat, the secondary seal will seat." The NASA people had gotten the answer they were looking for.

Although there were 16 NASA people present at this teleconference, it is worth noting that all of them were employees of Huntsville. No one was there from Houston, or the Cape, or Headquarters in Washington. Within an hour of the meeting, Lawrence Mulloy telephoned Arnold Aldrich, the shuttle program manager (whose actual title is Manager, National Space Transportation System Program Office, Johnson Space Center, Texas). Mulloy reported that bad weather out at sea would delay recovery of the SRBs when they parachuted into the ocean, but no mention was made of possible joint problems or the discussions with Thiokol. Later, Chairman Rogers asked Mulloy why not. "At that time, and I still consider today, that was a Level III issue. . . . We work many problems . . . that never got communicated to Mr. Aldrich. . . . It was clearly a Level III issue that had been resolved."

In NASA parlance, Level I is a Flight Readiness review at the highest level, chaired by someone from NASA Headquarters, usually Jesse Moore, Associate Administrator for Space Flight. The Level II review is conducted by Mr. Aldrich, the program manager. Level III is at the major component level, in this case the SRBs, with authority residing in Huntsville. What Mr. Mulloy was saying was that he didn't consider the joint problem significant enough to bump it up one level in NASA's decision-making pyramid. He did, however, mention it to his boss, the Director of the Huntsville Center, Dr. William Lucas. Lucas later told the Rogers Commission that had he known the extent of the Thiokol engineers' concern, he would have objected to the launch.

Arnold Aldrich testified to the Commission that not only had he not been told about the teleconference on the evening of January 27, but that he had several other complaints about how the system had been operating: " . . . critical problems should be brought forward to Level II and not only to Level II, but through myself to Level I. The second breakdown in communications, however, and one that I personally am concerned about, is the situation of the variety of reviews that were conducted last summer. . . . "

Mr. Aldrich went on to say, in effect, that NASA Headquarters and Huntsville had been bypassing him and dealing directly with such problems as how to test new joint configurations and how to get the money for additional SRB casings. He also complained that Huntsville had not shared completely with his office in Houston the data from the various flights, so that he had not understood the full extent of seal erosion and blow-by problems.

But if Aldrich (Level II) had been bypassed, did not Moore (Level I) have the full facts about SRB joints? Apparently not. The Huntsville officials claimed that they and Thiokol together had presented a comprehensive briefing at NASA Headquarters the summer before, on August 19, 1985. Unfortunately, Jesse Moore was not present at this session, which was attended by his Deputy, Michael

Weeks (the same Mike Weeks who was McDonnell's chief of design during
the Mercury competition). Mike says that he summarized for Jesse the major
points of the briefing, but Moore says he doesn't remember that. At any rate,
Bill Rogers gathered together four key officials and put the question to them:
"Did any of you gentlemen prior to launch know about the objections of Thiokol
to the launch?"

Mr. Richard G. Smith (Director of the Kennedy Space Center): "I did not."

Mr. James A. Thomas (Launch Director, Kennedy Space Center): "No, sir."

Mr. Arnold Aldrich (Shuttle Program Director, Johnson Space Center):
"I did not."

Mr. Jesse Moore (Associate Administrator for Space Flight, NASA Headquarters): "I did not."

The decision to launch the Challenger was "flawed," concluded the Rogers
Commission. "If the decision-makers had known all the facts, it is highly unlikely
that they would have decided to launch . . . on January 28, 1986."

The Commission took a broad interpretation of its charter and commented
not only on what *did* cause the Challenger accident, but also on what *might*
cause future accidents. They came up with a fairly long list of worries, some
involving hardware and some procedures.

The Commission very rightly considers ascent a critical phase of a shuttle
flight, and frets over the crew's inability to escape under various circumstances:
"Crew survival during ascent rests on the following assumptions: (1) The Solid
Rocket Boosters will work from ignition to planned separation. (2) If more than
one main engine fails, the crew must be able to survive a water landing."

In other words, both SRBs must operate properly for the full 2 minutes and
8 seconds of their use. Analysis has shown that if the crew attempts to separate
from the SRBs while they are still burning, the orbiter will hang up on its aft
attach points and pitch violently, with probable loss of the vehicle and crew.
The Commission says that "new ideas and technologies should be examined"
to see if there isn't some way to shut down both SRBs gently if something has
gone wrong.

If two of the liquid engines fail, the orbiter will probably have to ditch in
the ocean—a crash landing that almost certainly would kill everyone on board,
because of the heavy weight and high gliding speed of the orbiter. But without
modifications the crew would have no choice but to glide to their destruction,
with several minutes to think about the inevitable crash.

Some people think that something similar happened to the Challenger crew,
that they survived the initial explosion and spent the next 2 minutes and 45 seconds conscious of the fact they were tumbling and spinning toward a certain
death below. Two months after the Rogers Commission finished its work, NASA
released additional information concerning the astronauts' "personal egress air
packs," or PEAPs.

Each crew member has such an air supply connected to his or her helmet, for use in ground emergencies. Four PEAPs were recovered from the ocean floor, and analysis showed that three of them had been activated manually. The contents of the PEAPs were partially depleted, indicating either that their owners were breathing or that PEAP air was leaking past helmet seals. However, I doubt that any of the astronauts were conscious for long.

The explosion took place at approximately 47,000 feet, and the shuttle's upward velocity carried the crew compartment up to a maximum altitude of about 65,000 feet. Exposed to these altitudes, the crew's time of useful consciousness would have been only around 15 seconds, even if they had been breathing PEAP air. That is the time it takes blood (in this case, unoxygenated) to circulate from the lungs to the left side of the heart and then to the brain, which cannot remain conscious without oxygen. The PEAP, unlike a pressure suit, cannot exert sufficient pressure to oxygenate the blood at altitudes above 50,000 feet.

All this assumes that the disintegration of the orbiter ruptured the crew compartment and caused it to depressurize. If the crew compartment had retained its pressure, the crew could have breathed cabin air and it is unlikely that they would have activated three PEAPs. Therefore my conclusion is that the crew was unconscious for most of the 165 seconds it took their cabin to reach a most certainly fatal impact with the ocean. Of course, it is possible that they might have regained consciousness as they descended into dense air.

The Rogers Commission thought about the crew escape problem, and considered ejection seats and a separable crew compartment. It heard testimony from several astronauts and reluctantly agreed with Bob Crippen: "I don't know of an escape system that would have saved the crew from the particular incident that we just went through (the Challenger accident)." However, the Commission believes that the crew should have a means of escaping the orbiter "in controlled, gliding flight." *Controlled* and *gliding*, the two key words, not while still riding the SRBs or tumbling out of control. And that is one of the things NASA is working on now. Perhaps simple parachutes with a modified escape hatch will be sufficient, but more likely some kind of personal rocket harness, to extract each crew member, will be required.

In addition to the ascent, the Commission found the descent and landing fertile soil to till. First the selection of a suitable landing site, especially with regard to the weather at Cape Canaveral: "Once the shuttle performs the deorbit burn, it is going to land approximately 60 minutes later: there is no way to return to orbit, and there is no option to select another landing site. This means that the weather forecaster must analyze the landing site weather nearly 1½ hours in advance of landing, and that the forecast must be accurate. Unfortunately, the Florida weather is particularly difficult to forecast at certain times of year. In the spring and summer, thunderstorms build and dissipate quickly and unpredictably. Early morning fog is also very difficult to predict if the forecast must be made in the hour before sunrise. In contrast, the stable weather patterns at Edwards make the forecaster's job much easier."

Bob Crippen agrees: "I don't think the astronaut office would disagree with the premise that you are much safer landing at Edwards." The report goes on: "Shuttle program officials must recognize that Edwards is a permanent, essential part of the program. The cost associated with regular, scheduled landing and turnaround operations at Edwards is thus a necessary program cost." And a big one at that, for the orbiter has to be hoisted onto the back of a Boeing 747 and flown back to the Cape, a process that is not only expensive and adds 6 days to turnaround time, but which exposes the orbiter to additional damage as well.

Once a suitable runway has been selected, the heavyweight glider must be brought to a safe stop on it. The Commission found fault with the orbiter's tires and brakes, considering them marginal. The tires are very good, but most tires don't have to handle the impact of a 240,000-pound vehicle hitting a runway at terrific speeds. One instant they are motionless, and the next instant they are turning at 250 mph. The tires have a 34-ply rating, using 16 cords, but after use they still show excessive wear, especially when crosswind landings are involved, something that worries the Commission " . . . because loss of a single tire could cause loss of control and subsequent loss of vehicle and crew."

The brakes are also suspect. They have been damaged on most flights, and they must be applied very gingerly. The Commission agrees with John Young, who says: "We don't believe that astronauts or pilots should be able to break the brakes." Some runways that might be needed in an emergency, such as Dakar in Africa, are short and have poor overruns.

Other pieces of hardware disturbed the Commission. The most obvious are the three liquid engines, the space shuttle main engines. Some SSME components are wearing out more rapidly than predicted, the Commission noted: "The high-pressure fuel turbopump is limited to six flights before overhaul . . . the life-limiting items on the high-pressure pumps are the turbine blades, impellers, seals, and bearings. Rocketdyne has identified cracked turbine blades in the high-pressure pumps as a primary concern."

These pumps have always scared the hell out of me. They operate at very high speeds—35,000 rpm—and as they spin they produce enough centrifugal force that should one let go, that is, disintegrate, fragments fly out radially with sufficient force to penetrate adjacent machinery. And guess what is adjacent? The other two SSMEs in a neat little cluster. The Challenger explosion occurred just 15 seconds after the three SSMEs had throttled up from 65% to 104% of rated thrust, and my first reaction was that one of them had let go, and flying debris had penetrated the External Tank. It couldn't, I thought, have been one of the good old dumb SRBs that let go.

The Commission also pointed out another potential catastrophic failure point. Hydrogen and oxygen are fed from the ET to the three engines through two 17-inch diameter lines. In each line there is a valve, one controlling the hydrogen flow, the other the oxygen. "Each of the disconnect valves has two flappers that close off the flow of the liquid hydrogen and oxygen when the ET separates from the orbiter. An inadvertent closure by any one of the four flappers during normal engine operation would cause a catastrophe due to rupture of the supply line and/or tank.''

The Rogers Commission interviewed more than 160 individuals and amassed 12,000 pages of transcript. It examined over 100,000 pages of documents. It worked with a staff of 15 experienced investigators gathered from various parts of the government. Thousands of NASA and contractor personnel were involved in supporting roles, but only one NASA employee was a member of the Commission. Some old NASA hands grumbled that it was a witch hunt by outsiders trying to make themselves look good, but as far as I'm concerned, it is a very nice piece of work—comprehensive, timely, and of great value to NASA as it tries to glue itself back together in the aftermath of the Challenger explosion.

The culmination of the Rogers Commission's report is a listing of nine recommendations:

I. The SRB joint and seal must be fixed, and *really* fixed: ''The integrity of the structure and of the seals of all joints should be not less than that of the case walls. . . .''

II. The shuttle program management should be altered, to better define the authority of the program manager, and to use the operational experience of astronauts in management positions. A safety panel should be formed, reporting to the program manager.

III. NASA and its contractors should review all hazardous items and identify those that have to be fixed before the shuttle flies again.

IV. NASA must elevate its safety organization, and have it report directly to the NASA Administrator.

V. Communications must be improved between Huntsville and the rest of NASA. Flight readiness reviews must be updated, and attended by the flight crew commander.

VI. Landing safety (weather, tires, brakes) must be improved.

VII. In regard to launch abort and crew escape, every effort should be made to increase the range of conditions under which an emergency landing can be successfully made, and to provide a crew escape system during controlled, gliding flight.

VIII. NASA must establish a flight rate that is consistent with its resources, and avoid the ''relentless pressure'' that came from the STS being this country's principal space launcher.

IX. Maintenance safeguards must be established to prevent ''cannibalizing'' one orbiter to provide parts for another, and to ensure that an inspection plan is rigorously followed.

Concluding Thought:

''The Commission urges that NASA continue to receive the support of the Administration and the nation. The Agency constitutes a national resource that plays a critical role in space exploration and development. It also provides a symbol of national pride and technological leadership. The Commission applauds NASA's spectacular achievements of the past and anticipates impressive achievements to come. The findings and recommendations presented in this report are intended to contribute to the future NASA successes that the nation both expects and requires as the 21st century approaches.''

Reaction to the Rogers report has been almost universally favorable, although some on Capitol Hill thought a few heads should roll. Rogers, a former Attorney General, told both Senate and House panels that he did not think prosecution would be successful, nor would it be in the national interest. "I hope it doesn't happen," he said, and it hasn't. Huntsville was the only part of NASA that did not cooperate fully with his Commission, he said, and Congressman James Scheuer, a Democrat from New York, strongly criticized what he called a "cult of arrogance" there.

I know from my own NASA days that Huntsville has always been a little different from the other centers. Never a part of the old NACA gang, it traces its heritage back through the Saturn series all the way to Peenemünde, and has been derisively referred to as "Hunsville" because of von Braun and his team. But my own experience is that von Braun was anything but a Teutonic authoritarian. He was a warm and friendly man, gregarious, open to new ideas, and, above all, a communicator. He was also a fine engineer. I cannot imagine von Braun sitting on a problem like the O-rings.

It is true that NASA managers have prided themselves on being decisive and, in my opinion, one of NASA's strengths has been its ability to decentralize, to push decisions down the pyramid. But at the same time, the old NASA had a freer flow of information. You kept your boss informed, even though you were not calling on him to make a decision. Problems were allowed, indeed encouraged, to bubble up through the system, not get cut off at Level IV or III or any other level. I don't remember any such numbering system. Jim Webb would have known, somehow. He might not have known the difference between a tang and a clevis, but he would have known that one of his contractors was out there waving a distress flag. His people would have told him.

Perhaps it was past success at Huntsville that caused a change in attitude there. Undeniably, the Saturn series of rockets was more successful than any reasonable person would have expected, and the Huntsville people had every reason to be proud of their record. In an organization, healthy pride can come very close to destructive arrogance. The attitude that "we'll handle our own problems, thank you" can be taken a step too far. It's all just human nature, I suppose, but reading some of the Rogers Commission testimony makes me feel that recently Huntsville has become a castle whose occupants have pulled up the drawbridge, leaving the moat full of serpents.

Max Faget remembers Huntsville as an organization of competent engineers led by hard-nosed program managers who were, nevertheless, willing to openly discuss their problems. However, they insisted on solving such problems by their own methods and standards which were, in every respect, thorough and conservative. If they have since been overly autocratic, he believes it has only been that way "for the last 3 or 4 years."

When queried, Lawrence Mulloy of NASA doesn't shed much light on the situation, telling the *Washington Post* that the Commission seemed to understand the joints better than the responsible engineers because "I wasn't smart enough, the people who advised me weren't smart enough, the contractor wasn't smart enough . . . the people who reviewed my activities weren't smart enough." Before the House Science and Technology Committee he blamed "group think" for the launch decision. Mulloy has retired from NASA. Jesse Moore, who has been reassigned within NASA, said, "It has been an especially difficult year for me and it is beginning to have an adverse effect and take its toll on my family." William Lucas, Director of Huntsville, has retired. At Thiokol, Roger Boisjoly has left the company after an extended period of sick leave.

It's a sad turn of events for this young agency that had become accustomed to a string of successes. It had no contingency plan for failure, and in the wake of the biggest crisis in its history, NASA has floundered. James Beggs, generally regarded as the strongest NASA administrator since Jim Webb, relinquished his job just 2 months before Challenger to defend in a lawsuit brought in an unrelated matter. He left a vacuum that became filled with internal dissension. Huntsville has angered the Cape by raising the possibility that Challenger's O-rings were damaged during assembly there. Huntsville and Houston have squabbled about the division of future labor between the two centers. Outside critics have jumped in, especially those who prefer unmanned scientific forays into space. "What I see out there," said astronaut Henry Hartsfield, "reminds me of a sick chicken in a chicken yard," with everyone pecking at it.

The astronauts with whom I have discussed the situation don't express surprise or anger that seven of their friends were blown up, but rather a feeling of sadness and disillusionment, the latter usually stated as "management let us down." They know full well the risks of their profession, and before any spaceflight there are doubts and fears about various bits and pieces of the machine, but this first, inflight fatal accident seems so unnecessary. The matter of the O-rings, they say, was periodically presented as a minor problem awaiting an improved design, but in the meantime, no sweat. Yet I can certainly sympathize with the Huntsville managers who noted that flight after flight, the secondary O-rings had remained intact. I might have made the same decision they did.

Commission member Richard Feynman was smart enough to know better than that. "For 10 years they have discussed the problem and didn't do anything about it. . . . The guys who know something about what the world is really like are at the lowest levels of the organizations, and the ones who know how to influence other people . . . they're at the top. . . . For a successful technology, reality must take precedence over public relations, for nature cannot be fooled.'' I don't agree that PR replaced reality—I think he was off the track on that—but ''nature cannot be fooled,'' that is the heart of the matter.

But hasn't NASA been through all this once before, after the Grissom-White-Chaffee fire? Yes, but there are at least two important differences this time. One is the enormity of the Challenger catastrophe: *inflight, seven dead, on television, the first teacher in space.* What a lesson. Second, at the time of the Apollo fire in 1967 the clock was still ticking toward the end of the decade, and NASA didn't have time for too much back-biting or soul-searching. NASA and North American snarled at each other briefly, but then it was fix the problem and get on with it.

This time there is no such imperative. Take your time. Consider all possibilities. Fix everything. Take no chances. Astronaut Joe Allen likens NASA to a person. In 1967 it was a young, resilient teenager. Now it's middle-aged, and recovering from a serious wound takes a lot longer. He also faults what he sees as tentative, timid leadership, and points out that delay itself imposes an element of danger. No sensible pilot is too eager to fly a ''hangar queen,'' an airplane that has languished on the ground for a long time.

Chief Astronaut John Young also misses the sense of urgency that permeated NASA after the Apollo fire. He decries a ''business as usual'' attitude on the part of management, which he perceives as having too many layers, filled by people who don't have the proper background. One of his predecessors, Deke Slayton, said he used to be able to talk directly to his Center Director and the NASA Administrator, but John has to go through several intermediate layers. John has always been a prolific memo writer, and he almost always writes in what I call a ''gloom and doom'' style. I kidded him about this. ''You know me,'' he answered, ''I'm more pessimistic than most, but it isn't pessimism, it's just being realistic.'' Had he been expecting an accident somewhere along the line? ''I thought it would be during landing,'' he replied, but added in the next breath: ''But my stomach sinks every time I see a launch.'' Would he fly the shuttle again? ''Maybe,'' he answered, depending on the quality of the fixes.

In one of his first memos after Challenger, with a little more bombast than usual, John addressed the subject of safety: ''By whatever management method it takes, we must make Flight Safety First. People being responsible for making Flight Safety First when the launch schedule is First cannot possibly make Flight Safety First no matter what they say. . . . If we do not consider Flight Safety First all the time at all levels of NASA, this machinery and this program will NOT make it. If the management system is not continuously self-assessing with respect to Flight Safety of the inherently hazardous business that we are in, it will NOT last. If the management system is not big enough to STOP the space shuttle program whenever necessary to make Flight Safety corrections, it will

NOT survive and neither will our three space shuttles or their crews. . . . Our space machinery is not airline machinery.''

NASA has recently announced a plan for resuming shuttle operations, with the next flight scheduled for 1988. The SRBs will have new joints, featuring added insulation, a third O-ring, and a metal grip to reduce flexing. The President has approved building a replacement orbiter, at a cost of $2.8 billion, but it won't be ready to fly until 1989 at the earliest. Pilots have a superstition that accidents come in threes, and that seems accurate in the aftermath of Challenger, depending on how the count is made.

Less than 3 months after Challenger, one of the Air Force's heavy-duty launch vehicles, the Titan III, blew up. Then the next month, on NASA's first launch attempt since Challenger, an old, reliable Delta tumbled out of control. It was the worst string of failures in many years, and a national embarrassment.

With both the shuttle and the Titan III grounded, suddenly the nation found itself without any heavy lift capability. For years the Air Force had been complaining that it was not prudent to put all the nation's big eggs in the shuttle basket, and they had been allowed to buy a few Titan IIIs. But the pendulum has swung, and NASA officials who, a short time ago, were championing the shuttle, are now extolling the virtues of a balance between not only manned and unmanned rockets, but manned versus robotics operations in space.

The Air Force is procuring a new version of the Titan, the Titan IV. The White House has announced that, to spur the development of commercial launch vehicles, the orbiter in the future will avoid most commercial payloads, carrying only those that are shuttle-unique or those that have national security or foreign policy implications.

The problem is that there is a growing backlog of payloads waiting to be launched. NASA had planned over a dozen shuttle flights in both 1986 and 1987, and its commercial customers are now left high and dry. The French Ariane has more business than it can handle, and the Russians are offering their powerful Proton booster to the world market at the bargain price of $24 million per launch. The Japanese are beginning to develop an all-new heavyweight launcher, called the H-2. Even the Chinese are wading in, with their '' Long March'' rocket, and indications are that some U.S. commercial satellites will soon be launched by it.

In the meantime, NASA is working to make the shuttle a safer vehicle, and the STS a safer system, not just with regard to O-rings but any other parts that are suspect. Whether NASA is working with a sufficient sense of urgency or whether it is "business as usual" depends on to whom you listen. Admiral Richard Truly, a former astronaut who inherited from Jesse Moore the Headquarters hotseat as Associate Administrator for Space Flight, says that certain critical functions involved with joint redesign and testing are being scheduled 7 days a week, 24 hours a day.

On the other hand, my own observation is that other parts of NASA are discouraged, morale is low, and a feeling of lassitude pervades the organization. They need a lift, they need to get flying again. But even that may not restore the sense of excitement that I remember from Apollo days. After all, the shuttle was designed to make access to space routine, and whatever becomes routine tends to become dull as well. It's just not possible to keep people on the edge of their chairs forever. Sooner or later they are going to relax, sit back, and put their feet up.

NASA was badly shaken by Challenger, and jarred out of any such feeling of complacency, but—in the absence of a compelling goal such as the moon— how do you find and keep the best people, and keep them in a high state of dedication and concentration forever? NASA's work force has been growing older, and in recent years the Agency has had trouble attracting the very best college graduates, partly because of noncompetitive government salaries and partly because the magic is disappearing.

I suppose organizations are only as good as their people, and NASA still has many of the finest, but organizations also seem to assume a character of their own. To me, walking the halls of a NASA installation was always different from visiting the Department of Commerce, or State, not that I have anything against those organizations. But NASA was new, and people scurried about with zest, with a youthful spring in their step. Now NASA seems pretty much like other old-timers, a mature bureaucracy, a bit set in its ways, shuffling, not dancing, through austere times. Its arteries are hardening a bit.

I believe, however, that it can spring back quickly. I just hope it and the administration don't overreact to Challenger. NASA simply cannot guarantee 100% safety. If it tries, it will never get flying again. If this period of grounding drags on too long, more and more engineers will conjure up more and more objections or improvements. Top management may become afraid to ignore any warnings, no matter how ill-founded. The safety profession seems to attract more than its share of zealots and it takes an experienced and gutsy leader to know when their precautions are becoming excessive.

I recall that just a matter of days before Apollo 11 left earth, I received a dire warning from a NASA consultant I knew. The gist of it was that the surface of the moon might be littered with small nodules of pure metallic elements. Brought back as dirt on the boots of Armstrong and Aldrin, this material might spontaneously ignite when the Lunar Module was repressurized with pure oxygen. Would I please do something about this? I discussed the matter briefly with some people whose judgment I trusted, and then promptly forgot it. There comes a time to stop the handwringing and go fly. When it is grounded NASA is just another government bureaucracy, and we have plenty of those already.

In my study I have three framed group photographs. One is of student test pilots, one is of experienced fighter test pilots, and one is of fledgling astronauts. Of the 50 men in these photos, taken roughly 25 years ago, 11 are dead. Two died of heart attacks, one in an automobile accident, seven were killed in aircraft, and one (Roger Chaffee) in a spacecraft. Although admittedly this is a pitiably small sample size, from a statistical point of view, I nonetheless conclude from it that flying aircraft is disproportionately dangerous, not spaceflight. (Unfortunately astronauts get a double whammy, because they are exposed not only to spaceflight, but also to the full spectrum of accidents and diseases.) Seeking a larger sample size, I note that in 1 year cigarette smoking is suspected of killing over 300,000 Americans, and at least 40,000 of us are sure to die on the nation's highways. I am not trying to diminish the horror of any *preventable* accident, especially Challenger's, but merely to put the number seven into some perspective. Should NASA be judged against a standard of perfection?

NASA will have been grounded for nearly 3 years following the Challenger accident. This interval has been chaotic as various commercial, military, and scientific payloads have searched for alternative launch vehicles. The Air Force has been nursing the Titan III back to health, building a Titan IV, and launching small payloads aboard the Delta rocket. The scientific community is bringing up the rear, as usual, and remains grounded. The shuttle, which was to be all things to all customers, has for the moment become a national embarrassment.

When NASA does resume shuttle flights, it will do so with more caution and more attention to scheduling at reasonable intervals, so that the people at Cape Canaveral can work under less pressure in refurbishing the orbiter between flights. Four flights in 1988, ten in 1989, eleven in 1990: NASA feels it can sustain a rate of approximately one launch a month, whereas a few years back it was contemplating one per week. Another change is that NASA will no longer be allowed to launch commercial payloads unless they have been tailor-made to fit the orbiter or it is considered to be in the best foreign policy interests of the country to do so.

One of the 1989 flights will carry a most interesting payload, one that should whet the public's interest in space science as never before. It is called the Hubble Space Telescope, and it is designed to look out into the far corners of the universe, to see events that took place almost at the beginning of time, shortly after the "Big Bang" that created the universe and its subsequent expansion that continues today.

When we look through an earth-based telescope at the rings of Saturn, we are not seeing them as they are now, but as they were 1½ hours ago, when the light from them began its 1 billion-mile voyage to our eyeballs. The more distant the object, the more important this time lapse becomes. In the case of Saturn we can be fairly certain that the 90-minute transit time is of little consequence, that the fundamental fact of Saturn and its rings will not change appreciably over such a short span of time.

This is not so in the case of far-distant objects. Here we have a time history of the turmoil taking place in the violent expansion of our universe, as it bubbles and boils, eddies and swirls. In the process are created and destroyed such weird objects as pulsars, quasars, and black holes. In a single generation we have

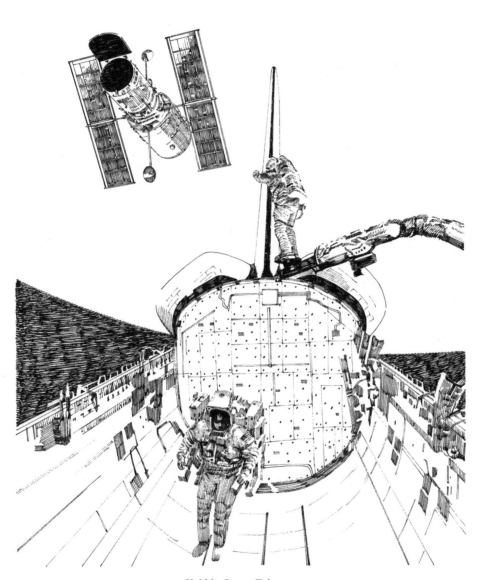

Hubble Space Telescope

learned more about the universe than all previous generations, yet the Hubble by itself offers the promise of expanding by an order of magnitude the amount of information we receive.

The Big Bang is believed to have taken place between 15 and 20 billion years ago. Today we can see out to a distance of less than 2 billion light-years, but with the Hubble we should be able to see 12 or 15 billion, into immense new regions of the universe, regions so far away that when their light reaches the Hubble we will know what was taking place shortly after the beginning of time.

Astronomers should, for example, be able to learn much about quasars, far-distant objects that despite their relatively small size give off more energy than some entire galaxies produce. Closer in, Hubble will be able to observe near stars and see whether or not planets circle them, a fundamental step in looking for worlds similar to ours and capable of sustaining our kind of carbon-based life.

How will the Hubble be able to do all these things? The trick is that the shuttle will release it in a 300-mile orbit, high above the earth's atmosphere. It is the atmosphere that frustrates astronomers, because it blocks over 90% of the energy coming in from the stars. Not only that, the atmosphere doesn't keep steady, and its wobbling causes star images to twinkle, or scintillate. The result is that even the largest earth-based telescopes, such as the 200-inch behemoth on Mount Palomar, have difficulty getting crisp, focused photographs. The Hubble is much smaller, only 94 inches in diameter, but it can look directly into the far reaches of space without the limiting and annoying filter of the atmosphere. It will be able to see seven times as far as Mount Palomar's telescope, and will allow scientists to view 350 times the volume of space they do now.

The Hubble Space Telescope will also be a ''permanent'' observatory, that is, NASA intends to keep it operating indefinitely. It is designed to be serviced in orbit by visiting astronauts. Hubble is a cylinder about 44 feet long and 14 feet in diameter, weighing 13 tons. When it needs servicing, the shuttle will rendezvous with it and pull it into the payload bay using the remote manipulator arm. Astronauts can then perform their maintenance or repair work from a platform in the payload bay. Its batteries will need to be replaced every 3 years or so, and its solar panels and scientific instruments at longer intervals. This remarkable machine is appropriately named in honor of Dr. Edwin Hubble, an early American astronomer and Rhodes Scholar who was the first to offer clear evidence of an expanding universe.

The Hubble's successful launch will, I feel, be the most important piece of work NASA has done in recent years, and one that I hope will herald the Agency's return to the forefront of science and exploration. It will be an institutional as well as scientific milestone.

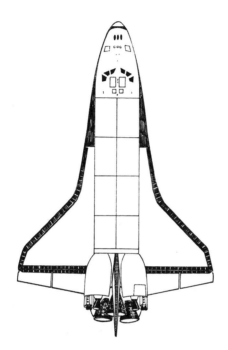

Shuttle in flight

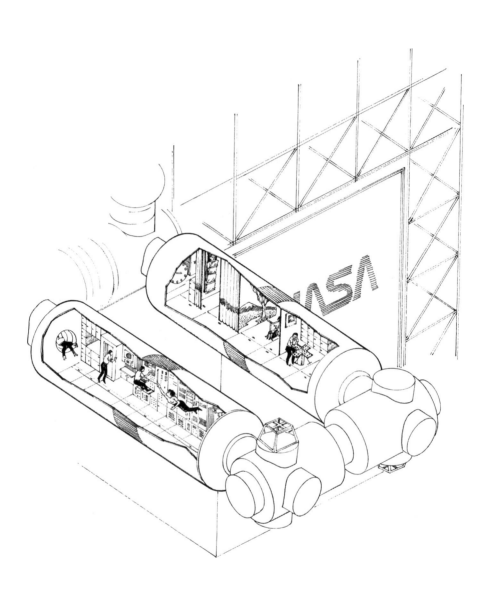

7

Ad Inexplorata

In 1945, Arthur C. Clarke, the science fiction writer, originated the concept of the communications satellite. From the very beginning of space exploration he has been a wise commentator, watching the space machines develop, and listening to their proponents extol their virtues. When the shuttle first appeared it was described as a workhorse, the DC-3 of the Space Age, a comparison to the venerable Douglas transport airplane designed in the 1930s and still flying today. Clarke continually refers to the shuttle as the DC-1½, putting it accurately in its place in the astronautical sequence. But the space DC-3, 707, and 747 are yet to come.

Today a quiet but intense debate is going on within NASA, the White House, and the Congress as to how the nation should proceed beyond the DC-1½. A space station has been approved, but it does not seem to be generating much enthusiasm, perhaps because the Soviets already have one in orbit. Yet a permanent space station appears to be the logical next step for NASA. It will be assembled bit by piece using the shuttle and its extravehicular astronauts. Several dozen shuttle trips will be required to complete the basic station, and the shuttle will also be used to swap crews and replenish expendable supplies, such as fluids, oxygen, food, and raw material for experiments.

The station will be a multi-purpose facility; a national laboratory in space that will also be a permanent observatory; a facility for servicing, assembling, and manufacturing; a transportation node (by that NASA means a central location); a storage depot; and a staging base. As far back as 1952, Wernher von Braun wrote that "Development of the space station is as inevitable as the rising sun," yet 35 years later only the Soviet Union has launched a permanent orbiting outpost.

Skylab, of course, was a space station of sorts, but was permitted to return to Earth after less than 6 months of human occupancy. The space station that von Braun had in mind was an ornate affair, an orbiting torus not unlike the one in Stanley Kubrick's movie *2001*. The station NASA proposes is crude by comparison, a truss-work Tinkertoy that can change its size and shape as new purposes for it are discovered, and new modules added.

The NASA station will not be small. Its basic structure consists of dual keels, each 360 feet long and connected by upper and lower booms that, with solar panels and antennas extended, could be over 500 feet wide. NASA will start with a scaled-down version. One plan, for example, would begin with neither keel but with one of the transverse booms. Within seven shuttle flights this boom could be assembled, complete with solar panels and two pressurized modules, one a habitat for the astronauts doing the construction and the other a laboratory. Within 3 years the entire dual keel configuration would be completed.

The station will probably be an international venture, and the European Space Agency and the Japanese seem eager to become full-fledged partners in the venture. NASA is semi-eager. On the one hand, it badly needs financial assistance, because the station price tag keeps growing and may reach $20 billion. On the other hand, managing a large team of American contractors is complicated enough without the added headaches involved in melding foreign cultures and interests into the process. The Japanese, for example, want to provide a laboratory module, which in a way is skimming the cream off the top. They want the United States to take care of the basic stuff, like supplying power and a place for the crew to live, while they control the design and perhaps operation of that part of the station most apt to produce commercial benefits. The Europeans want to contribute a manned module, to be named Columbus, that could also fly free of the station, giving them a bit of independence for reasons of prestige and perhaps substance.

In the wake of the Challenger accident, NASA observers seem to perceive a mandate to offer criticisms of the space station. NASA employees, past and present, have also joined in with suggested improvements. Former astronaut Owen Garriott thinks that the 15-foot diameter of the shuttle payload bay will result in modules that are of marginal volume. He wants a heavy-lift booster to replace the shuttle as a station launcher, so that the diameter of a module could be extended to 25 feet and its length to 90 feet. Owen argues that such an unmanned booster would save money, compared to using the shuttle, and that the redesign of the space station would be beneficial, similar to the evolution from wet workshop to dry workshop during the Skylab program.

Other astronauts are concerned about the fact that the station has no "lifeboat" but relies on a "safe haven" concept, wherein the crew must react to a fire

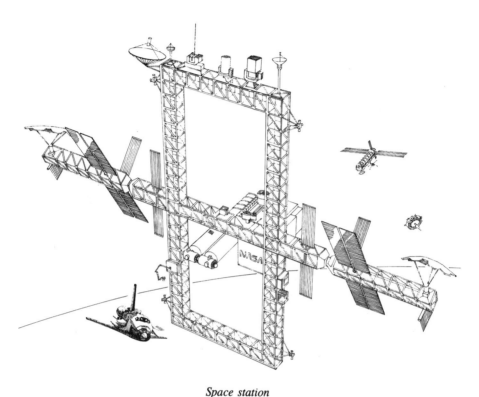

Space station

or other emergency by retreating to an isolated corner of the station and waiting for a shuttle to come rescue them. In response, NASA is studying the design of a crew emergency return vehicle.

To James Van Allen, the notion of investing a large chunk of American cash in the station is just one more example of NASA's silly and wasteful preoccupation with sending people out to do a robot's job. The University of Iowa physicist, for whom the earth's two radiation belts are named, has flayed NASA's manned program from the very beginning: "Apart from serving the spirit of adventure, there is little reason for sending people into space." According to some people, apart from a spirit of adventure, there is little reason for savoring life on earth. The *New York Times* has also sniffed at the space shuttle from the beginning, calling it "an expensive yawn in space."

The Defense Department's reaction to NASA's plans has been mixed. First the Pentagon said it had no requirement for a military man in space, and no interest in the station. Later it hedged its bets and said it might want to do military experiments on board, and warned NASA to make sure its foreign partners understood that.

Consequently, the precise design and nature of the space station is far from clear. Its supporters claim that important scientific and commercial breakthroughs will be made, once a proper laboratory environment for experiments is provided. What experiments will be performed in the station's laboratory? There is a long list of possibilities, but if experimentation in weightlessness is similar to laboratory work on earth, the most useful discoveries will be ones that just pop out as research unfolds, rather than those that are consciously pursued. Skipping over the possibility of such celestial serendipity, there are three ideas for processing materials that appear promising.

The first is called electrophoresis. Suspended particles are caused to move through a fluid in response to an electrical current. In earth laboratories electrophoresis is used in biological research, but it only works with minute quantities of fluid because gravity causes sedimentation and convection currents, two very powerful distorting effects. In weightlessness a fluid can be separated into its components more rapidly and in much greater quantity. Apollo and shuttle experiments have succeeded in producing pharmaceutical products that are ultra-pure. The next step is to produce them in commercial quantity at competitive prices. It is estimated that by the year 2000, between 1 and 2 billion vaccinations a year will be administered. Space manufacturing could increase the purity, and hence safety, of many of these vaccines. Other substances that might be produced aboard the station are *beta cells* to control diabetes, *erythropoietin* to treat anemia, and *interferon*, which affects the body's immune system.

Second is the manufacture of crystals for the electronics industry. Gallium arsenide, for example, is a material that allows electricity to flow through it at higher speed than most materials. Gallium arsenide transistors are faster than computer chips made from silicon, the material commonly used today. Gallium arsenide also resists radiation better and remains stable over a wider range of temperature. One U.S. aerospace firm wants to fly a series of experiments to form gallium arsenide crystals in a process called directional solidification, involving precisely controlled temperatures that first melt and then solidify the material. In weightlessness, during the molten phase, the internal crystalline structure of the gallium arsenide is precisely aligned to prevent or correct any structural imperfections. In this manner crystals of greater size and purity could be formed.

The third idea is to create new metal alloys in space. Many interesting combinations cannot be homogeneously mixed on earth because the lighter elements float to the top or the heavier sink to the bottom, whichever way you look at it. In weightlessness, not only is there perfect mixing, but gas bubbles can be distributed throughout the alloy, reducing its weight. It is likely that a metallurgical laboratory and small foundry will be important components of the station.

In this book so far, I have referred to conditions aboard a spacecraft as being weightless, or in zero gravity. I have done so because when I was an astronaut we always called it that. However, these terms are only approximately correct, and need closer examination when considering a laboratory in space. The problem is that aboard any spacecraft there are disturbances. The crew moves around and bumps into things. The attitude control system can never align the vehicle to perfection, but fires jets that cause the spacecraft to sway back and forth on either side of the desired direction. Gases are vented overboard, and cause minute accelerations. The upper atmosphere may exert a tiny drag effect. All these actions disturb the purity of the environment, so that G, or acceleration measured on board, is not precisely zero, but more like 1/1000 or 1/10,000 of the acceleration due to gravity we experience on earth, which is 32 feet per second, per second. I don't include all these little numbers to nitpick, but to explain why the term *micro-gravity* has been coined, and why it is more accurate than zero gravity, or weightlessness. The closer to zero-G a laboratory can come, the better it is for the undisturbed, pure creation of crystals, pharmaceuticals, and alloys during their liquid or molten phase. Don't jiggle the laboratory, the soufflé will fall.

The disturbances, or accelerations, inside a space laboratory depend in part on where it is located relative to the rest of the machinery. Near the center of mass of the station, a laboratory would experience smaller acceleration effects than if it were swinging out on the end of a long boom. Scientists want pure zero-G on one end of the scale, but they are also interested in a centrifuge that could spin at various speeds, producing G levels between zero and one. In this manner they could study, for the first time, the effect of gravity on various materials and organisms in a precisely controlled manner.

Of course the organism the medical doctors are most interested in studying is the human. The information they have received from Skylab and the Soviet space stations has only whetted their appetites. If astronauts and cosmonauts lose bone density at the rate of 0.5% each month while they are in space, how long can they be kept up there? How about a trip to Mars that might take a couple of years? The present plan is to rotate a station crew of six to eight people at 90-day intervals, but I'm sure once it gets past its shakedown phase, people will be kept up there for longer and longer intervals. Today cosmonaut Yuri Romanenko holds the record—11 months—having spent nearly all of 1987 aboard the Space Station Mir.

What would it be like to spend a year aboard the station? NASA says the station will include private crew quarters, a combination galley and wardroom, a health maintenance and exercise area, and hygiene facilities. All that in the habitat module, plus, of course, an opportunity to visit the laboratory and perhaps Japanese and European modules. For privacy and recreation, each crew member will have a compartment containing a sleep restraint, an audiovisual entertainment center, writing desk, storage space, and controls for lighting and ventilation. The galley/wardroom will be the social center as well as the place for food storage and preparation. The health and exercise area will have exercise machines and medical diagnostic equipment.

We can conclude from the Mercury, Gemini, and Apollo programs that highly motivated individuals are willing and able to perform at an elevated level under far from ideal conditions for brief periods of time, but it is equally clear from Skylab that as things get more routine and the weeks drag on, psychological factors become increasingly important. After an 8-day flight to the moon, astronauts could expect a ticker-tape parade. After a year aboard the space station, they might get a half-priced drink during Happy Hour. To quote a NASA study, " . . . when humans are subjected to adverse conditions for long durations, on a routine basis, and in the absence of substantial reward or recognition, performance degradation *must* be expected." Yet if the station is to be successful, crew productivity must be maintained at a high level.

In seeking solutions to this potential problem, NASA has studied Antarctic research stations, remote military outposts, nuclear submarines, and undersea habitats—in addition, of course, to the Skylab and shuttle designs. The Agency has even looked to racing yachts, commercial fishing vessels, and offshore oil platforms for lessons to be learned. And lessons to be avoided, too. For example, during one winter at an Antarctic research station the average alcohol consumption consisted of 9 ounces of distilled liquor plus 5.4 beers per man per day!

The closest non-space analog to the proposed station seems to be Sealab II, a 12×57-foot cylinder located 200 feet below the surface of the ocean. The U.S. Navy organized three teams of 10 men each, composed of military and civilian divers, scientists, and salvage experts, and each team spent 10 days aboard. As a result of these studies, a long list of design considerations emerged. Chief concerns center around: sleep, noise control, safety, clothing, exercise, recreation, medical support, personal hygiene, food selection and preparation, outside communications, waste disposal, on-board training, habitat aesthetics, privacy and personal space, and group interaction.

Naturally the crew selection process will to a large extent affect that last factor, group interaction. Once the crew has been assembled, its command structure and social organization will be equally important. A rigid, military-style chain of command probably will be the simplest and safest system, especially considering that inflight emergencies may require a swift and highly disciplined reaction. On the other hand, aboard a research facility the crew will tend to divide itself into two factions: those doing the research and those responsible for the operation of the station itself. Scientists feel, with considerable justification, that the whole purpose of the station is to learn by conducting tests and experiments, that their work is paramount, and therefore they should be in charge. I think most NASA employees, certainly astronauts, take the position that the captain of the ship must be in command, even though the vessel is carrying some very important passengers.

As we have seen from the Skylab flights, especially Pete Conrad's, the personality of the commander is all-important. Writing of the voyage of Thor Heyerdahl's papyrus raft, *Ra*, in the *New York Times*, Walter Sullivan reported that "The patterns of alliance and hostility fluctuated . . . although Mr. Heyerdahl always retained his position of leadership and good relations with all. A commanding personality, in such a situation, is extremely important . . . and Mr. Heyerdahl well fulfilled that role."

Without a strong leader, and perhaps even with one, trivial issues become exaggerated, and nerve endings get rubbed raw. A typical incident at an Antarctic research facility as reported by a Navy physician: "Cook's at it again. He's moody, definitely emotionally immature. Threw a lemon pie and cookies all over the galley the other day, then went to his room for a couple of days and wouldn't come out." This type of extreme withdrawal, or "cocooning," could have disastrous effects if permitted in a small space station. Not to mention the cookie crumbs floating all over the place and getting inhaled into crew members' lungs. My own attempt to relieve tension derives from the advice of a psychiatrist who said there are two cardinal rules: (1) Don't sweat the small stuff, and (2) It's all small stuff.

In earth orbit, human frailties can be relatively easily tolerated. If things get too bad, the earth is close at hand. In 1985, the London *Daily Telegraph* reported that one cosmonaut commander had to return to earth 2 months into a much longer mission. He was, according to his flight engineer, a "bundle of nerves." A trip to the moon, an 8-day round-trip, doesn't present much more of a problem than orbiting the earth. But how about far out, where the planets of our solar system beckon? There things will get rough indeed, and groups of travelers are going to have to be selected and trained with the greatest care. "Group dynamics" will no longer be psycho-babble, but a matter of life and death.

Our neighborhood planets are Mercury and Venus, closer to the sun than we are, and then those farther away: Mars, Jupiter, Saturn, Uranus, Neptune, and Pluto. Between Mars and Jupiter is a belt of asteroids, the dividing line between the four Earth-like inner planets and the outer gaseous giants. NASA has sent unmanned spacecraft to, or to the vicinity of, all of them except Neptune and Pluto, and Neptune will be visited in 1989. Poor little Pluto, whose orbit is cocked at an angle to all the rest (the others are in what is called the Plane of the Ecliptic), may not see a terrestrial machine for a long time to come.

Mercury, closest to the sun, looks a lot like our moon. Mariner 10 flew to within 600 miles of it in 1973, and photos show a dead planet, with a daytime surface temperature of 800°F —hot enough to melt lead. Like the other inner planets, Mercury has a metallic core surrounded by an outer shell of silicate material. In the past all the inner

Mercury

planets were alive with earthquakes and volcanos, and bombarded by meteorites, but Mercury has changed little in the past 3 billion years. It is only 40% larger in diameter than our moon, and it has no atmosphere. To land on Mercury and return to Earth, very powerful rockets would be required to overcome the gravitational pull of the nearby sun.

Venus, closest to the Earth, has long been known as the evening star and is, next to the moon, the brightest object in the night sky. Almost as large as Earth, Venus has a thick atmosphere composed mostly of carbon dioxide, but with a little bit of nitrogen, oxygen, and water vapor. Venus' yellow-white clouds form an opaque screen, but although we cannot see its surface we know that, like Mercury, it is hellishly hot, and has a surface pressure nearly 100 times that of Earth. The upper atmosphere does not look homogeneous, but shows lighter and darker swirls and waves that take 4 days to circle the planet.

Venus

Next comes the most beautiful planet, our Earth, 93 million miles from the sun, a distance we refer to as one astronomical unit (a.u.). Its satellite, the moon, circles it at a distance that varies slightly, but averages 238,000 miles. As viewed from the moon, the Earth is the most beautiful object I have ever seen. I could watch it with fascination for hours on end: tiny, but bright and shiny, blue and white, a place of oceans and clouds, a solitary and exquisite outpost floating through the velvet black of space. The Earth takes 365¼ days to make one revolution around the sun, rotating at the rate of one turn every 24 hours. The seasons are caused by the fact that its axis is tilted slightly. The tides are caused by the gravitational pull of the moon. The Earth is at an ideal distance from the sun; much closer and we would sizzle, much farther away, freeze.

Earth

What turn of celestial luck or design put us here I do not know, but a close-up glimpse of our neighbor the moon presents an arresting contrast, and convinces me that we humans have the ultimate prize of this solar system, and perhaps of the entire universe.

Beyond the Earth lurk some really fascinating places. Mars, the red planet, next out from the sun, has intrigued me all my life. We have visited it with Mariner and Viking unmanned spacecraft and placed instruments gently on its surface. Slightly over half the diameter of the Earth, it is quite like the Earth in some respects, vastly different in others. Similarities include the fact that it orbits the sun in the same plane we do, and it turns on its axis once every 24 hours and 39 minutes. Its axis is tilted at 25° to the Plane of the Ecliptic instead of our 23°. It takes 687 days to travel completely around the sun. Thus, although its year is almost twice as long as ours, it has four seasons and nearly the same length of day.

Its topography is extraordinarily varied, with mountains three times as high as Mount Everest and an immense chasm that puts the Grand Canyon to shame. Early astronomers thought that through their telescopes they could see straight canals crisscrossing the Martian surface, but none exist. There are, however, a number of meandering channels that seem to be the result of floods that occurred over a billion years ago. That water has long since vanished, and today Mars has no liquid on its surface. It does have ice caps on both poles, composed both of water ice and "dry" ice—frozen carbon dioxide. Also, water probably exists in the form of permafrost beneath the Martian soil.

Mars

Mars has an atmosphere, but compared to ours it is very thin—less than 1% of our sea level barometric pressure. It is composed almost entirely of carbon dioxide, with traces of oxygen, nitrogen, and argon. The atmosphere is thick enough to produce high winds and towering dust clouds. One gigantic storm in 1971 featured hurricane force winds that raged for months, totally obscuring the surface. Thin white clouds are fairly common in the Martian atmosphere. The Viking spacecraft landed in areas that look similar both to the moon and to Earth's deserts. An abundance of rocks and boulders litters the flat surface, composed of reddish, sandy soil. As seen from the surface, the sky is a delicate pink.

Nothing lives on Mars, at least not as far as the two Viking landers could tell. They scooped up soil samples and analyzed them chemically. The soil has a composition much like weathered lava, an indication of the strong influence volcanism has had in forming the Martian surface. But no microbes or other living creatures could be induced to grow in the soil pulled into Viking's test chambers, an indication that no life exists at the two locations NASA chose to investigate. Underneath the polar ice caps, who knows? Astronauts walking the surface might find more interesting places than did the Viking landers. Astronauts on Mars would weigh just 38% of their Earth weight, and it would take somewhere between 5 and 20 minutes for their comments to travel back to Earth, depending on the alignment of the two planets.

Two satellites orbit Mars. To an astronaut on the Martian surface the larger—Phobos—would appear one-third the size of the Earth's moon, while its companion—Deimos—would look like a very bright star. Up close the two look like dark potatoes that have been partially nibbled away. Phobos is slightly over 15 miles long and Deimos about 10. Phobos is heavily pockmarked by impact craters; Deimos is slightly smoother. They are named after sons of the Greek war god: Fear and Terror. They are important because, from a gravitational point of view, they are easier to reach than Mars itself. An expedition could land on either of them without having to expend the extra fuel required to fight Mars' gravity all the way down to its surface and then back up again.

Launched in 1972, Pioneer 10 became the first man-made object to exit the solar system and in the process flew through the asteroid belt. Some people at NASA feared that Pioneer 10 would be destroyed by a collision with one of these tiny planets that form an orbiting swarm between Mars and Jupiter.

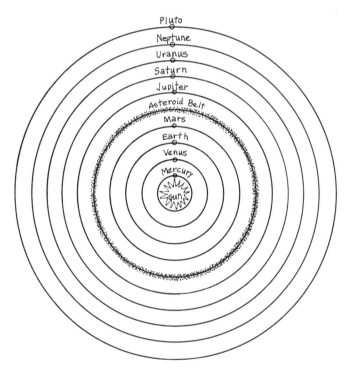

The solar system

A few of them are large (Ceres has a diameter of nearly 500 miles), but even a collision with one as small as a golf ball would have put Pioneer 10 out of service because of the high velocity of impact.

It is impossible to count all the asteroids, but if they were assembled into one planet astronomers estimate that it would have a diameter one-eighth that of Earth. Some think that the asteroids were once such a body that was destroyed by collision or by the gravitational force of nearby Jupiter. Today most experts believe that the asteroids are remnants of the original nebular material that formed the solar system, remnants that were prevented by Jupiter's gravity from accreting completely into a planet. The asteroids are believed to be the source of most meteorites and, if so, are composed of such materials as glass, lava, and hunks of nickel-iron. Many asteroids, instead of staying put between Mars and Jupiter, have orbits that zing in toward the sun. Some, like Icarus, penetrate even closer to the sun than Mercury. In 1958, Icarus came within 4 million miles of the Earth (a very near miss on a cosmic scale). When can we expect a collision? While a direct hit by an asteroid as big as Ceres would probably destroy our planet, collisions with the tiny ones we call meteorites are helpful because some of them are 4.5 billion years old and tell us a bit about what the solar system was like in its early, formative stage.

Jupiter, 5 a.u. out from the sun, is the giant of the solar system, with a diameter 11 times that of Earth and a mass over 300 times ours. Jupiter weighs over twice as much as all the other planets combined. We have sent Pioneer and Voyager probes to Jupiter and passed within 26,000 miles of it. Jupiter is sur-

rounded by intense radiation belts, in some places a million times as strong as Earth's Van Allen belts. Jupiter's magnetic field, in similar fashion, is much more powerful than Earth's, and a terrestrial compass would point at Jupiter's South, not North, Pole. Jupiter radiates over twice as much energy as it receives from the sun.

Jupiter is composed primarily of hydrogen, but a strange form of hydrogen. In its upper layers it is the gas we know, but as the pressure builds up

Jupiter

down below, the hydrogen first condenses into a liquid and then is transformed into a highly compressed, metallic state. The surface of Jupiter looks like a marble I had as a kid, a prized agate "shooter" with bands of brownish red. Jupiter's bands eddy and swirl, and one gigantic storm, called "the eye of Jupiter," forms a permanent red spot about the size of the planet Earth.

Perhaps of even greater interest than Jupiter itself are the satellites that ring it. There are at least 16 of them in a confusion of inner, outer, and intersecting orbits. The four largest ones were discovered by Galileo in 1610. They have names from Greek mythology that to me are lyrical, beautiful, almost magical: Io, Europa, Ganymede, Callisto. I think I would like to visit them. Each is different from the others, and together they present a bewildering collection. Io is red-orange, Europa yellowish, Ganymede variegated, and Callisto brown. Io has active volcanoes spewing out sulfur and sodium. Europa is crisscrossed with dark lines that apparently delineate surface fractures. Ganymede looks somewhat like our moon but also features curved grooves that look like the plow marks left by some celestial farmer. Callisto has been peppered by meteorites, including perhaps the solar system's largest, whose aftermath is a series of rings circling a shallow icy basin. Europa is the smallest (2,000 miles in diameter), Ganymede the largest (3,300 miles).

A billion miles from Earth, Saturn is the planet with the rings—ethereal rings of amazing symmetry and grace. The Voyager photographs of the rings and the shadows they cast on Saturn have a surreal quality, as if a technical illustrator had used his sharpest pencils and most precise instruments to trace an endless series of delicate ellipses. Saturn is second in size only to Jupiter and, like Jupiter, is a ball of hydrogen, but mostly in liquid form. It is butterscotch in color and shows thin striations of pink, tan, and brown that circle its axis of rotation.

Saturn

Saturn has at least 17 moons, the largest of which is Titan, almost the size of Mercury. To me Titan is a special place in this solar system, second only to Earth in its ability to nurture life. There are certainly things wrong with it, from that perspective, but it does have an atmosphere, thicker than ours, and it does contain nitrogen, methane, and other gases that were present in our own atmosphere billions of years ago. Its surface may be as cold as minus 300°F, inhospitable indeed for life, and it is probably a mixture of ice, rock, and metals. But it also includes a dusting of various organic molecules, including hydrocarbons and single carbon-nitrogen compounds. Who knows what kind of subterranean lakes might exist, and in them, at warmer temperatures, closer to Titan's hot core, what kind of life might have evolved? I remember the fairly recent discovery, deep in our ocean, of tube worms clustered in water warmed by volcanic vents. These creatures exist by using a most strange metabolism based on the consumption of hydrogen sulfide, and independent of sunlight.

Uranus, Neptune, and Pluto form the far outposts of our solar system. There is some tidiness in their placement, in that Saturn is roughly 10 a.u. from the sun and these three are located at 20, 30, and 40 a.u., respectively. To make one orbit around the sun takes Uranus 84 years, Neptune 165 years, and Pluto 248 years. Uranus is different from all the other planets in that it orbits on its side instead of upright. At one point in its orbit, the sun is shining directly on its North Pole and 180° later on the South Pole. It takes 2 hours and 45 minutes for a radio signal to travel from Uranus to Earth.

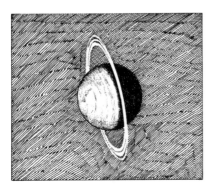
Uranus

Although tiny compared to Jupiter and Saturn, Uranus still is 14 times more massive than Earth. It, too, has rings and moons, but its rings seem insignificant compared to Saturn's and its moons much more remote than Jupiter's. Uranus looks like a blue-green billiard ball, the color coming from the methane in its upper atmosphere, where winds howl at 500 mph. Voyager visited Uranus in late January 1986, and returned remarkable photographs, especially of the moon Miranda, but they received scant public attention because they arrived at the time of the shuttle Challenger's explosion.

Invisible to the naked eye, Neptune was discovered in 1846 after a bit of nifty detective work. Astronomers noted that something was pulling at Uranus, causing it to speed up at some positions in its orbit and to slow down at others. By calculating where in the sky this hidden magnet might be, the astronomers trained their telescopes on that spot and—voila—Neptune!

Neptune

A Voyager spacecraft will visit Neptune in August 1989, and until then our knowledge of it will remain rather sketchy. Neptune is about the same size as Uranus and may or may not have rings. Its upper atmosphere is composed of methane and perhaps ammonia and hydrogen. It has one huge moon, Triton, and a tiny one, Nereid. Triton is odd in that it orbits backwards, that is, opposite to the direction in which its planet spins.

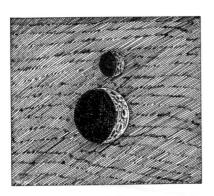
Pluto and Charon

Pluto and its traveling companion Charon are also odd. The rest of the planets behave themselves and stay in the plane of the ecliptic, like peas on a plate. But not Pluto. Its orbit is tilted at an angle to all the others. It is smaller than our moon and very cold. Some astronomers think Pluto may not be a true planet but a moon escaped from Neptune's orbit. Others think it and Charon, nearly a matched pair, are simply remnants from the solar system's formation some 4.6 billion years ago. At any rate, because of its remote location it may be generations before we find out much more about Pluto.

Will we ever visit these places—Mars, Titan, or even beyond? In gauging the future one is tempted to say what we *can* do, not what we *will* do. The *can do* of space travel is complex enough, involving enormous technological challenges, but the *will do* is far tougher. It encompasses our technological and scientific skill, economic capacity, political will, and perhaps even a philosophical or religious component. The conventional wisdom is that we tend to overestimate what we are going to do over the short haul and underestimate our long-term progress. In my own case I know the former is true—my desk is seldom clear at the end of the day, as I have promised myself. But the long-term visions that science fiction writers and others have had—how about them? Are they possible, or practical, and if so, when?

Tom Paine has a clear and comprehensive view of the future. A former NASA

Thomas Paine

Administrator, in 1985 he chaired the National Commission on Space, whose charter was to write a 20-year blueprint for NASA's future. Tom's group decided the best way to plan for 20 years was to consider what might happen in 50, and their 200-page report is a compendium of all that is possible over that period. Boiled down to its essentials, the Paine Commission grouped their findings into three broad categories: (1) "Advancing our understanding of our planet, our solar system, and the universe"; (2) "Exploring, prospecting, and settling the solar system"; and (3) "Stimulating space enterprise for the direct benefit of people on Earth."

To accomplish these goals economically, they believe that the nation must develop two supporting strategies, "advancing technology across a broad spectrum . . ." and "creating . . . systems. . . to provide low cost access to the space frontier."

To reach these goals, the Commission outlined 12 milestones along the path to Mars:

- Initial operation of a permanent space station;
- Initial operation of dramatically lower cost transport vehicles to and from low Earth orbit for cargo and passengers;
- Addition of modular transfer vehicles capable of moving cargoes and people from low Earth orbit to any destination in the inner Solar System;

- A spaceport in low Earth orbit;
- Operation of an initial lunar outpost and pilot production of rocket propellant;
- Initial operation of a nuclear electric vehicle for high-energy missions to the outer planets;
- First shipment of shielding mass from the moon;
- Deployment of a spaceport in lunar orbit to support expanding human operations on the moon;
- Initial operation of an Earth-Mars transportation system for robotic precursor missions to Mars;
- First flight of a cycling spaceship to open continuing passenger transport between Earth orbit and Mars orbit;
- Human exploration and prospecting from astronaut outposts on Phobos, Deimos, and Mars; and
- Start-up of the first Martian resource development base to provide oxygen, water, food, construction materials, and rocket propellants.

The cost of all this? The Commission is "confident that the long-range agenda we recommend can be carried out within reasonable civilian space budgets." The U.S. economy will continue to expand at an average rate of 2.4% annually for the next 50 years, the Commission believes, and—compared to the gross national product—NASA's appropriation could stay less than half that required during the peak Apollo years.

The Commission report was published in May 1986, in the wake of the Challenger accident. To many this seemed most unfortunate timing, and the Paine Commission's work "disappeared without a trace," as one old Washington hand put it.

I asked Tom Paine whether his report hadn't come at the worst possible moment. "Of course it did," he replied, but then quickly added: "But in another sense—the Dickensian sense—it was the best of times as well as the worst. This country *must* face up to its future in space, and Challenger brought that to the fore. Our report lies there as a direction to form one position of a national debate."

Dr. Paine seems to worry not so much about space *per se*, but space as a descriptor, or manifestation, of the kind of country the United States is and chooses to be in the future. "Today western civilization is on trial. We have moved a long way from Victorian values and are trying to build a new society. The U.S.A. is the model for the world in this, but it seems the U.S.A. has lost its way. . . . Space can be a key question in this. . . . We have to get the nation back on track and NASA can make the U.S.A. the leading technological nation, to lay the foundation. We need to export high-tech products, not pass laws restricting the export of technology. Space can be 'a lotus flower emerging from a cesspool,' something with real purpose to sustain public support."

But who is going to pay for all this? "Today we have a trillion-dollar budget.

In that context it's trivial whether the NASA budget is 6, 8, or 10 billion dollars. What's important is: what kind of nation is America going to be? Last year they were arguing about an *extra* $30 billion for the Department of Defense. That's four to five times the whole NASA budget!'' At this point Tom was getting red in the face and waving his arms. ''It's much more important to understand what areas this country is going to emphasize. . . . What careers are our youngsters going to pursue, will they be lawyers or entertainers or study science and technology? . . . The U.S. may decide not to lead on the space frontier but then if we turn our back, as Toynbee pointed out, nations rise and fall. The space frontier is the fundamental American challenge . . . we must associate our wealth and power with positive values. We cannot turn our backs on technology. Today we are at a crossroads. My biggest fear is that NASA won't understand its role.''

I asked Tom about the importance of leadership in setting goals for NASA. ''We need leadership. Without it would be like having a Marshall Plan without General Marshall or the occupation of Japan without Douglas MacArthur. The Soviets understand this. At first they were flabbergasted by the impact of Sputnik, but they have learned the lesson almost better than we. . . . We need a leader to stand up. . . . Beyond the station, the Reagan goal should be permanently occupied and supported outposts on the moon and Mars, and a transportation system through the inner solar system to support them. Kids will respond to this kind of vision. 'Got to take chemistry,' their teacher tells them. 'No, it smells.' 'Got to take it.' 'Okay.' 'Physics, computers too.' 'Okay.' ''

I've always had a warm spot in my heart for Tom Paine. He was Executive Officer of the submarine *Pom-pom* during World War II combat in the Pacific, and I think he likened the Apollo Command Module to a miniature submarine. A couple of days before the flight of Apollo 11 he flew from Washington to Florida to talk to Neil, Buzz, and me. A visit from the NASA Administrator was no small thing, and we were attentive. His message was that we were to take no unnecessary chances. But then he added the one ingredient that made it most likely we would really follow this standard advice. If you don't like the way things look, he said, come on home and I'll guarantee you three the next flight to try again. That took a lot of pressure off us, especially Neil, and I really appreciated it. It was a wise and thoughtful move.

Back in my office after talking to Tom, I pulled out some numbers from the 1987 federal budget. In round numbers the Defense Department's appropriation was $300 billion, the Department of Health and Human Services (including Social Security) got $200 billion, and NASA $10 billion (including $2 billion extra to start building a replacement for Challenger). In other words, the Defense Department spends the entire NASA budget every 12 days and it takes Health and Human Services 2 or 3 weeks to go through $10 billion.

My own criticism of the Paine Commission report may be unfair, but it is that the report is so complete, so well balanced, and offers such a comprehensive view that it will be of little use to citizens or government officials as a mechanism for getting NASA started again. We need a banner to wave, and a 12-point program is too much to embroider on it. "Less is more," as the architects say.

I would wave a very small flag, but wave it vigorously. On it would be printed "MARS." The quest for Mars would pull in its wake most things NASA is trying, in a fragmented fashion, to do today, plus create the climate for American enterprise and leadership that Tom Paine seeks. A space station would be a necessary precursor, as would the development of lower cost transportation. A return to the moon might be involved. But the country would have a destination, a focus for the whole range of technologies required to rebuild American preeminence.

Why Mars? I'll admit the place holds a special fascination for me, and has all my life. Inhospitable though it may be, it's the closest thing to a second home that we have in this solar system, with available oxygen and water. An expedition there is within the realm of the possible by the end of this century, with a price tag slightly less than that of the Apollo program.

However, I'm not recommending another headlong rush like Apollo's, but rather an acknowledgment that we humans will extend our domain out through the solar system, that Mars is the next logical destination, and that today the United States should make this commitment to exploration and begin laying the foundations for it, without setting a timetable or defining the cost of the first footstep on Mars. But we need to get started soon, or the world will pass us by.

The trip will take somewhere between 12 and 36 months. The path that consumes the least fuel is a half ellipse, tangent to the Earth's orbit at one end and Mars's orbit at the other. It takes 9 months to traverse this half ellipse, which is called a Hohmann transfer. Much faster trajectories can be worked out, but they gobble up lots more fuel and, as always, weight is a critical factor. Two Hohmanns require 18 months, but an added complication is the fact that the planets must be in proper alignment for such a transfer, and the crew may have to wait on the Martian surface for over a year for the right take-off time. To be on the safe side, let's call it a 3-year voyage.

Earth orbiting crews can be on the ground within hours of an emergency. In Apollo, return from the moon took less than 3 days. We thought that was a long, long time when we considered all the various parts that might break, but how about *a couple of years*? Humans and all their essential gear must be able to last that long, depending on

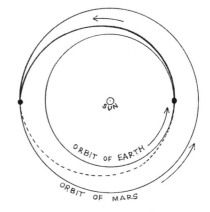

Hohmann transfer

where the trouble occurs. *Reliability* then becomes the key. This is the same quality that we prize in television sets and automobiles, the factor that seems to be giving the Japanese preeminence today. "Mars-quality" on a product will mean a new level of reliability has been achieved.

The human body will also have to endure with a reasonable safety margin. Columbus lost men on his voyages, but his adversaries—primarily malnutrition and disease—were probably more formidable than those facing Mars travelers. Certainly our ability to cope with disease has improved vastly over the past century, and infectious diseases are much less apt to occur inside an isolated spacecraft. Providing proper nutrition should pose no difficulties for the first Mars trips but, oddly enough, when self-sustaining colonies begin, it may very well become a problem. People talk of raising rabbits and growing strawberries on Mars, but we'll have to wait and see.

Two serious medical problems are with us today, but each has an engineering solution. First is radiation, severe enough to exceed a nuclear power plant worker's lifetime limit in a single Mars trip. Some type of radiation storm cellar (surrounded by water tanks?) will probably have to be provided as a refuge during peak solar activity. The second problem is bone demineralization in weightlessness. Unless some way is devised to stop it, or at least slow it down, a Mars spacecraft may have to be rotated to provide artificial gravity—an engineering nuisance, to say the least.

Then there are the inevitable human frailties, physical and mental. The capability to cope with simple broken bones, for instance, would certainly need to be part of the expedition's resources. On the other hand, even comparatively simple surgery would probably not be possible. The possibility of mental illness, or at the least aberrant behavior, is equally alarming. A fundamental consideration is the size and composition of the crew. Sexual activity, or the absence of it, will loom large in the minds of the crew over a 3-year period unless, like the Chinese emperors, we choose to create a cadre of eunuchs. The crew size of the initial voyages has yet to be defined. In 1952, Wernher von Braun proposed 70 people (10 ships, each with a crew of seven), but later pared it down to two ships and 12 people. That sounds about right to me. A dozen slots to fill . . . but with whom? Married couples? Half men, half women? All one sex? Or pick scientific and technical brains, and never mind the gender? Fascinating questions, with answers, perhaps, to be worked out aboard the space station.

One way of increasing safety is by redundancy. Von Braun and others have proposed two identical craft for the round-trip. A catastrophic failure in one would be handled by transferring the crew to the other. Of course the total weight—and hence cost of the expedition—would increase using this scheme. Once in the vicinity of Mars, a specialized landing craft (similar to Apollo's Lunar Module but using aerodynamic braking and parachutes in addition to retrorockets) will be required. Then, in order to traverse the Martian surface for reasonable distances, some sort of Rover would be advisable. These machines

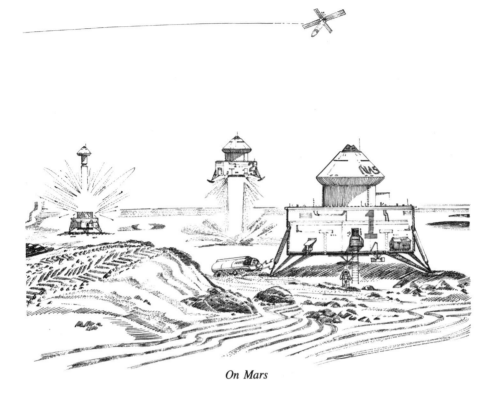

On Mars

will be launched separately and assembled in earth orbit, like the space station.

Designing and manufacturing all this hardware will present a challenge to American industry in almost all technical areas, such as materials, electronics, and quality control. Two particularly critical points stand out: propulsion and the environmental control system.

Apollo-type chemical propulsion using hydrogen and oxygen may suffice for the Mars mission, but it is less efficient than other systems on the drawing board. There are various nuclear-powered engines, ion drives, and even electromagnetic catapults. Ion drives can use a nuclear reactor as a power source, or perhaps even solar cells. The idea is to accelerate and expel from the engine a beam of ions—electrically charged atoms. In the process ("for every action there is an equal and opposite reaction") thrust is produced and the spacecraft gains speed. Ion engines are tiny compared to chemical rockets, but they can operate continuously for weeks or months instead of just minutes.

Up to now, the environmental control systems aboard our spacecraft have been "open loop," that is, certain substances are consumed and that is the end of it, with some waste products stored and some dumped overboard. In a closed loop or closed cycle, some fluids and gases will be recycled and used over and over again.

The average man needs 13.5 pounds per day to keep him going: 2.5 pounds of food, 9.2 pounds of water and 1.8 pounds of oxygen (the average woman uses less of everything, an argument in her favor that grows stronger as mission duration increases). The 13.5 pounds is converted by the body into waste products: 0.3 pounds solid, 11 pounds liquid (water) and 2.2 pounds gaseous (carbon dioxide).

In an open loop system, all 13.5 pounds are lost after one use, meaning that a crew of 12 in a 3-year mission would consume an absolute minimum of 180,000 pounds of food, water, and oxygen. That's a lot to haul up from the earth's surface, and it doesn't even include water for bathing or laundry. The loop must be closed and ways devised to recycle some of these substances just as the earth's ecological system does.

Reclaiming waste water is the beginning. Wash water, urine, and cabin humidity condensate can all be purified and used again for drinking. Electrolysis may also be used to convert some of the reclaimed water into oxygen. Carbon dioxide can be handled in a number of ways including feeding it and sunlight to plants and getting oxygen and food in return. Solid wastes can also be used as fertilizer to assist in this process. But whatever systems are adapted, they must be extraordinarily reliable, for the crew's life depends on them. If, halfway to Mars, the plants sicken and die, the way my houseplants do. . . . In all likelihood, during the early flights plants will be considered too risky because they depend on the complex behavior of various microorganisms, some of which must continue to function while others must remain quiescent.

Once on the surface of Mars there are new opportunities for closed-loop operation. Water can be obtained by dehumidification of the atmosphere and by extraction from the soil. At the poles, water may be abundantly available just by melting ice. The water can be electrolyzed to produce oxygen to breathe and hydrogen to react with atmospheric carbon dioxide to produce methane, a fuel suitable for Rovers, rockets, and heating purposes. Greenhouses can be constructed and plants grown in the Martian soil. Animals may be raised for protein. The soil can also provide building material, such as metals, glass, and cement. The total support of a dozen people would require greenhouses whose area would be about that of five football fields.

To some extent the first landing site will be selected with such considerations in mind, making it as easy as possible for the crew members to protect and feed themselves. But if an additional goal is exploration, the site must also be of particular interest to geologists and other scientists. Riverbeds, with their outcrops of layered rock, have a special appeal, as do the polar regions, where drilling below the icecap might reveal—who knows? It's possible that a long-range roving vehicle will permit a number of interesting spots to be examined in detail.

It is doubtful that we will discover life on Mars, but more likely that fossil evidence of previous life can be found in its rocks. On the other hand, it is certainly possible that Mars has never sustained life. In that case, scientists would like to know why. Carl Sagan has written, "But if Mars *is* lifeless, we have two planets, of virtually identical age, evolving next door to each other in the same solar system: life arises and proliferates on one, but not on the other. Why? This is the classic scientific circumstance of the experiment and the control . . . a better understanding of Mars leads to a better understanding of our own small planet."

The journey home from Mars will be much like the trip out: long hours of boredom, as the airline pilots say, but hopefully *not* punctuated with their moments of stark terror. The Martian rocks and soil will be suspect, even though no crew member has become ill over the months, and a stopover may be necessary at the space station (or even the moon) for laboratory analysis. As in the case of Apollo, the horrendous consequences of introducing alien pathogens into the earth's atmosphere would warrant the most conservative approach possible.

Naturally the United States has no special claim on Mars, and in fact does not even have a head start in a possible race there. Mars has fascinated people everywhere, and the Soviet literature reflects this as strongly as our own, although we are more familiar with names like Percival Lowell than Konstantin Tsiolkovsky. Since the 1960s the Soviets have used "Forward to Mars" as a slogan, and they have backed up their words with a series of unmanned probes. In 1988 they plan to send one to Phobos, to hover a few feet above it and blast the surface with a laser, and then analyze the resulting cloud of gas. Nor are the Soviets bashful about their plans to send people to Mars. In 1982, I heard their scientific attaché in Washington say that they would have a shuttle within 5 years and would send "men to Mars in 10 or 15 years." That would be between 1992 and 1997—more optimistic dates than most observers believe possible. However, 1992 does have a special significance for the Soviets because not only is it the 500th anniversary of Columbus' voyage but—more important to them—it is the 75th anniversary of the Bolshevik Revolution. They might very well decide to commemorate that with a trip to Mars, using hardware from the MIR space station.

There is great appeal to making an international venture of Mars's colonization, and momentum for the idea seems to be picking up. In this country, Carl Sagan has argued for it as a peaceful substitute for "Star Wars," and Senator Spark Matsunaga has devoted an entire book to the subject.

My own view is that it is certainly worth a try, but we should not design our own mission so that it will fail if the promised collaboration does not materialize. I'm skeptical about working with the Soviets, and I certainly don't believe they are going to drop their own Star Wars research, Mars or no Mars. The best way to ensure cooperation, I believe, is to have the clear capability to go it alone if necessary. But it would be a wonderful thing to have the Soviet Union and the United States actively working together on such an expedition, which should also involve assistance from European and Asian countries. Imagine

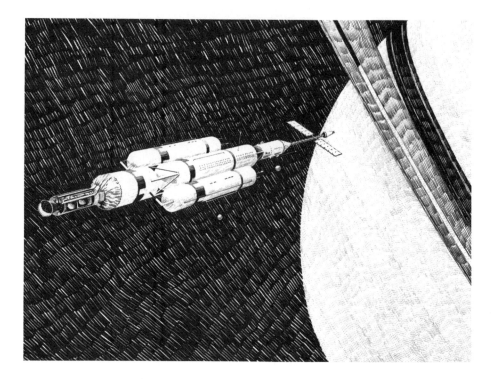

Starship

such a major undertaking, without weapons, 3 years of counting on each other for life-and-death matters. If it could be done in space, why not on earth?

I also hope that space can be used for more frivolous purposes. A resort hotel in orbit (an offshoot of the space station, perhaps?) would be immensely popular, enjoyable, and rewarding, if the price could be brought down to earth—so to speak. The view alone would make the trip worthwhile, and it might change the guests' attitude toward their planet in a fundamental way. The earth's borderless beauty is far more spectacular than photographs or words can convey, and a day or two in orbit would create a lifetime of unforgettable images. At least it has in my case. To most Americans the island of Taiwan means cheap electronics, but not to me. I see it glistening in the sunlight from 100 miles up, a verdant gardenia leaf floating in a shiny sea, indigo tinged with jade. In my memory it is a crisp and flawless picture.

Of course the weather in orbit is always nonexistent, or below you, or perfect—however you want to think of it. Never a day without sunshine, or a night without stars—fat, unblinking stars. Photographers, astronomers, geographers, or just those seeking a change: I guarantee that orbiting is the experience of a lifetime, a graceful and serene roller coaster ride around this magnificent planet of ours, and only 90 minutes per turn. Hotel guests might even return to earth determined to help make it the fairyland it appears from orbit.

Then there is the matter of weightlessness. Out the window is the feast of the eyes, but inside is for the entire body. It's a totally different world with unique opportunities for new sensations. No one is overweight here. All float with the same delicacy, aerial ballet dancers all. And lovemaking! I don't think any astronauts have yet been privileged to sample the ultimate use of weightlessness, but having no gravity to crush bodies together offers exquisite possibilities. A tiny pull here, a gentle touch there. A space *Kamasutra* remains for some lucky couple to write. Like swallows mating on the wing, no—far better than that—lovers in an orbiting hotel could explore their minds and bodies in ways simply not possible here on earth.

Procreation in space will probably be necessary if mankind ever ventures beyond this solar system. Places like Mars and Titan are but one grain of sand down the celestial beach, although they seem far to us because we move so slowly. The fastest humans have ever traveled is the 25,000 mph that Apollo astronauts needed to break earth's gravity and head out toward the moon. But at this speed it would take over 30 years just to cross our solar system.

To reach another sun and its planets we cannot think in terms of miles but must shift to light-years. Light, traveling at 186,000 miles per second, covers nearly 6 trillion miles in 1 year. According to Albert Einstein's theories, nothing can move faster than the speed of light. If Einstein is correct, and this cosmic speed limit must be observed, it will take over 4 years to reach Alpha Centauri, the nearest star. Nearly a decade for a round-trip, if we could travel at the speed of light, which we cannot yet do. At 25,000 mph it would take over 100,000 years to reach Alpha Centauri.

So far no one has been able to refute Einstein's special theory of relativity, although some believe that ways may be found to circumvent it. Writing shortly before his death, Charles Lindbergh contemplated the problem: ". . . now I find myself asking: what lies for man beyond solar system travel? . . . Is it remotely possible that we are approaching a stage in evolution when we can discover how to separate ourselves entirely from earthly life, to abandon our physical frameworks in order to extend both inwardly and outwardly through limitless dimensions of awareness? In future universal explorations, may we have no need for vehicles or matter? Is this the adventure opening to man beyond travel through solar-system space?''

Others theorize that there exists an entire separate universe in which the speed of light is the lower speed boundary, not the upper. In that domain particles called tachyons move faster than the speed of light and are unable to slow down to it, just as our light beams find it an impossible speed to reach. They are trapped in separate worlds, with the speed of light the impenetrable barrier between the two. But if somehow we could transfer back and forth between the two domains, then perhaps we could visit the far reaches of the universe. After all, at one time some aerodynamicists regarded the speed of sound as an impenetrable barrier.

But I will leave these speculations to physicists, or cosmologists, or philosophers. I'm content to think about what we might do within the confines of Mr. Einstein's limit. I don't think it unreasonable to expect that some day we will build a star ship, one that will depart earth never to return to this solar system. It will be easy, I believe, to fill this craft with volunteers, not nutty ones, but people of the very highest qualifications who will consider it a privilege to represent the human race as it expands to other worlds. After all, emigration is built into the human tradition. Our American ancestors had a pretty good notion of the "New World" before they left the old, but how about early Polynesians who departed their home island on a small raft for an unknown destination, with no intention of ever returning?

Besides, a star ship will have to be huge, sort of a self-contained township, and life aboard it will be more varied than that enjoyed by many people here on earth. Communicating with earth at the speed of light, the travelers will be far more worldly than a cloistered nun back home. Furthermore, they'll always have a new view out their windows, not the same old repetitive 365¼-day swing around the sun to which we earthlings are condemned. A year won't have the same meaning for them, nor will they have winter or summer unless they choose to create them artificially.

Where will they go? The Hubble Space Telescope and its successors will be used to reconnoiter, to look for suitable planetary systems. Its designers expect that Hubble will barely be able to make out planets surrounding nearby suns, but by the time a star ship is ready to depart, extensive surveys will have been conducted. We don't know how many planets are out there, but we can count stars and measure the distance to them. Just in our home galaxy, the Milky Way, there are 200 billion stars! Most are probably not suitable to sustain human life even if they have surrounding planets, but if the most pessimistic assumptions are made, there are still billions of stars that are the right size, age, and temperature—that are as capable as our own sun of nurturing life. Within a dozen light-years from Earth, in our stellar backyard, there are 15, including Sirius, the brightest of all stars as seen from Earth. Of these, Sirius and four others are binary, or double stars, offering the intriguing possibility of living on a planet with two suns in its sky, and a complex, intertwining orbit around them.

If there are suitable stars, in all likelihood there are habitable planets orbiting them, if our own solar system can be used as a model. In other words, there are probably billions of planets which, if not like Earth, at least possess characteristics hospitable to humans. Are they inhabited? It seems to me cosmic conceit to think that of all the billions of planets, Earth is the only one that has developed living beings. A more rational guess would put Earth toward the middle, meaning that somewhere in the universe there are millions of planets upon which life has existed longer than on Earth. But maybe not. Maybe we are the only one. Either way we should find out, first by listening, then by sending surrogate machines, and finally by going ourselves.

Will we ever do this, actually leave our home planet for good? I don't know when, but I am convinced we will. In a fundamental way Project Apollo was about leaving, the first move outward. Today the idea of a star ship seems preposterous because we don't have the necessary technology, but the pieces may be coming together faster than we realize. For example, our Air Force is already funding research on antimatter, a far-out concept in which a positively charged proton and a negatively charged antiproton collide, annihilating each other and in the process releasing a burst of pure energy. These particles can be produced in minute quantities today, and could be the key to driving a star ship close to the speed of light.

And a major fraction of the speed of light will be required, for a very odd reason. If a star ship departs at too slow a speed, its crew's descendants back on Earth will be discovering new efficiencies in propulsion, and will launch a more advanced machine a generation later that will pass the first!

Of course no one alive today will see these things happen, but that doesn't mean they won't. Our own generation is lucky in that we are the ones to explore the solar system for the first time, to see Apollo 8 carry three humans to escape velocity, to see Pioneer 10 sail out past Pluto. Every age has its dreams, and its symbols of those dreams. As a small boy mine were the racing planes of Jimmy Doolittle and Weddell Williams, and as an adult I've continued to look into the sky. Past planes now, past the moon even, I like to think about Mars and beyond. I don't want to feel a lid over my head, or the heads of my children. Their generation will find delectable planets, and want to visit them.

When the history of our galaxy is written, and for all any of us know it may already have been, if Earth gets mentioned at all, it won't be because its inhabitants visited their own moon. That first step, like a newborn's first cry, would be automatically assumed. What would be worth recording is what kind of civilization we earthlings created and whether or not we ventured out to other parts of the galaxy. Were we wanderers? Human history so far indicates that we are indeed. America in particular has a tradition of exploring, or expansion, of pushing back the frontier, across the mountains and plains, and then upward—with the Wright brothers, to the moon, and now beyond. It's human nature to do this—to stretch, to go, to see, to understand. Exploring space isn't an option, it's an imperative.

As for me, I just feel lucky to have been born in 1930. I grew up with biplanes and Buck Rogers, learned to fly in the early jet fighters, and hit my peak when moon rockets came along. Before I die, who knows what I may see? As Robert Goddard said, ''The dream of yesterday is the hope of today and the reality of tomorrow.''

GLOSSARY

ALSEP	Apollo Lunar Surface Experiment Package.
AMU	Astronaut maneuvering unit.
AOS	Acquisition of signal.
ASIS	Abort Sensing and Implementation System.
ATM	Apollo telescope mount.
ASTP	Apollo-Soyuz Test Project. Joining of United States and Soviet spacecraft in earth orbit.
BIGs	Biological isolation garments.
CAPCOM	Astronaut at Mission Control relaying advice and instructions to inflight space crews.
CM	Command Module. Mother ship.
CSM	Command and Service Module. Command Module and attached Service Module. Crew capacity: three.
ECS	Environmental control system regulating atmosphere inside spacecraft.
EOR	Earth Orbit Rendezvous Mode. Two independently launched spacecraft joined together while in earth orbit.
ET	External tank. Holds liquid fuel for second stage of STS.
EVA	Extravehicular activity in space, or "space walk."
HOUSTON	Manned Spacecraft Center.
HUNTSVILLE	Marshall Space Flight Center.
ISP	Specific impulse of rocket fuels.
LM	Lunar Module. Two-stage moon landing craft. Lower stage used for descent and then as launch pad for the upper, ascent stage. Crew capacity: two.
LOR	Lunar Orbit Rendezvous. Voyage to moon where spacecraft is a single unit until in moon orbit, and then separates into a mother ship (Command Module) and a landing craft. Astronauts use the landing craft to descend to moon, and return to mother ship. When mother ship departs for Earth the landing craft stays behind, parked in moon orbit.

LOS	Loss of signal.
LRL	Lunar Receiving Laboratory
MISS	Man in Space Soonest.
NACA	National Advisory Committee for Aeronautics.
NASA	National Aeronautics and Space Administration, or Space Agency.
PEAPs	Personal egress air packs.
RAND	Research and Development Corporation.
SM	Service Module. Storehouse attached to Command Module.
SPS	Service propulsion system. Command and Service Module rocket engine.
SRB	Solid rocket booster. Two SRBs form the first stage of the STS.
SSMEs	Space shuttle main engines (3).
STG	Space Task Group at Langley, in late 1961 changed to the Manned Spaceflight Center and moved to Houston, Texas.
STS	Space Transportation System. The shuttle.
SWIP	Super Weight Improvement Program.
TEI	Trans-earth injection. Start of voyage from moon orbit to earth.
TLI	Translunar injection. Start of voyage from earth orbit to moon.

U.S. MANNED SPACEFLIGHT LOG

MISSION	CREW	DATE	DURATION (DAYS:HOURS: MINUTES:SECONDS)
PROJECT MERCURY			
Mercury-Redstone 3	Shepard	May 5, 1961	00:00:15:22
Mercury-Redstone 4	Grissom	July 21, 1961	00:00:15:37
Mercury-Atlas 6	Glenn	Feb. 20, 1962	00:04:55:23
Mercury-Atlas 7	Carpenter	May 24, 1962	00:04:56:05
Mercury-Atlas 8	Schirra	Oct. 3, 1962	00:09:13:11
Mercury-Atlas 9	Cooper	May 15-16, 1963	01:10:19:49
GEMINI PROGRAM			
Gemini-Titan III	Grissom, Young	Mar. 23, 1965	00:04:53:00
Gemini-Titan IV	McDivitt, White	June 3-7, 1965	04:01:56:11
Gemini-Titan V	Cooper, Conrad	Aug. 21-29, 1965	07:22:55:14
Gemini-Titan VII	Borman, Lovell	Dec. 4-18, 1965	13:18:35:31
Gemini-Titan VI-A	Schirra, Stafford	Dec. 15-16, 1965	01:01:51:24
Gemini-Titan VIII	Armstrong, Scott	Mar. 16, 1966	00:10:41:26
Gemini-Titan IX-A	Stafford, Cernan	June 3-6, 1966	03:00:21:00
Gemini-Titan X	Young, Collins	July 18-21, 1966	02:22:46:39
Gemini-Titan XI	Conrad, Gordon	Sept. 12-15, 1966	02:23:17:08
Gemini-Titan XII	Lovell, Aldrin	Nov. 11-15, 1966	03:22:34:31

APOLLO PROGRAM

Apollo-Saturn 7	Schirra, Eisele, Cunningham	Oct. 11-22, 1968	10:20:09:03
Apollo-Saturn 8	Borman, Lovell, Anders	Dec. 21-27, 1968	06:03:00:42
Apollo-Saturn 9	McDivitt, Scott, Schweickart	Mar. 3-13, 1969	10:01:00:54
Apollo-Saturn 10	Stafford, Young, Cernan	May 18-26, 1969	08:00:03:23
Apollo-Saturn 11	Armstrong, Aldrin, Collins	July 16-24, 1969	08:03:18:35
Apollo-Saturn 12	Conrad, Gordon, Bean	Nov. 14-24, 1969	10:04:36:25
Apollo-Saturn 13	Lovell, Swigert, Haise	April 11-17, 1970	05:22:54:41
Apollo-Saturn 14	Shepard, Roosa, Mitchell	Jan. 31 to Feb. 8, 1971	09:00:01:57
Apollo-Saturn 15	Scott, Worden, Irwin	July 26 to Aug. 7, 1971	12:07:11:53
Apollo-Saturn 16	Young, Mattingly, Duke	April 16 to 27, 1972	11:01:51:05
Apollo-Saturn 17	Cernan, Evans, Schmitt	Dec. 7 to 19, 1972	12:13:51:59

SKYLAB PROGRAM

Skylab SL-2	Conrad, Kerwin, Weitz	May 25 to June 22, 1973	28:00:49:49
Skylab SL-3	Bean, Garriott, Lousma	July 28 to Sept. 25, 1973	59:11:09:04
Skylab SL-4	Carr, Gibson, Pogue	Nov. 16, 1973 to Feb. 8, 1974	84:01:15:32

APOLLO-SOYUZ TEST PROGRAM

(ASTP)	Stafford, Brand, Slayton	July 15 to 24, 1975	09:01:28:23

SPACE TRANSPORTATION SYSTEM

STS-1 (OFT)	Young, Crippen	April 12 to 14, 1981	02:06:20:53
STS-2 (OFT)	Engle, Truly	Nov. 12 to 14, 1981	02:06:13:12

STS-3 (OFT)	Lousma, Fullerton	March 22 to 30, 1982	08:00:04:49
STS-4 (OFT)	Mattingly, Hartsfield	June 27 to July 4, 1982	07:01:11:11
STS-5	Brand, Overmyer, Allen, Lenoir	Nov. 11 to 16, 1982	05:02:14:25
STS-6	Weitz, Bobko, Peterson, Musgrave	April 4 to 9, 1983	05:00:23:42
STS-7	Crippen, Hauck, Ride, Fabian, Thagard	June 18 to 24, 1983	06:02:23:59
STS-8	Truly, Brandenstein, D. Gardner, Bluford, W. Thornton	Aug. 30 to Sept. 5, 1983	06:01:08:40
STS-9	Young, Shaw, Garriott, Parker, Lichtenberg, Merbold	Nov. 28 to Dec. 8, 1983	09:07:47:24
41-B	Brand, Gibson, McCandless, McNair, Stewart	Feb. 3 to 11, 1984	07:23:15:55
41-C	Crippen, Scobee, van Hoften, G. Nelson, Hart	April 6 to 13, 1984	07:23:40:05
41-D	Hartsfield, Coats, Resnik, Hawley, Mullane, C. Walker	Aug. 30 to Sept. 5, 1984	06:00:57:00
41-G	Crippen, McBride, Ride, Sullivan, Leestma, Garneau, Scully-Power	Oct. 5-13, 1984	08:05:23:37
51-A	Hauck, D. Walker, D. Gardner, A. Fisher, Allen	Nov. 8-16, 1984	07:23:44:56
51-C	Mattingly, Shriver, Onizuka, Buchli, Payton	Jan. 24 to 27, 1985	03:01:33:27

51-D	Bobko, Williams, Seddon, Hoffman, Griggs, C. Walker, Garn	April 12 to 19, 1985	04:18:23:54
51-B	Overmyer, Gregory, Lind, Thagard, W. Thornton, van den Berg, Wang	April 29 to May 6, 1985	07:00:08:47
51-G	Brandenstein, Creighton, Lucid, Fabian, Nagel, Baudry, Al-Saud	June 17 to 24, 1985	07:01:39:00
51-F	Fullerton, Bridges, Musgrave, England, Henize, Acton, Bartoe	July 29 to Aug. 6, 1985	07:22:45:26
51-I	Engle, Covey, van Hoffen, Lounge, W. Fisher	Aug. 27 to Sept. 3, 1985	07:02:27:42
51-J	Bobko, Grabe, Hilmers, Stewart, Pailes	Oct. 3 to 7, 1985	04:01:14:38
61-A	Hartsfield, Nagel, Buchli, Bluford, Dunbar, Furrer, Messerschmid, Ockels	Oct. 30 to Nov. 6, 1985	07:00:44:51
61-B	Shaw, O'Connor, Cleave, Spring, Ross, Neri-Vela, C. Walker	Nov. 26 to Dec. 3, 1985	06:21:04:49
61-C	Gibson, Bolden, Chang-Diaz, Hawley, G. Nelson, Cenker, B. Nelson	Jan. 12 to 18, 1986	06:02:03:51
51-L	Scobee, Smith, Resnik, Onizuka, McNair, Jarvis, McAuliffe	Jan. 28, 1986	00:00:01:13

U.S. MAN-HOURS IN SPACE

Program	Mercury	Gemini	Apollo	Skylab	ASTP	STS
Man-hours in space	54	1,940	7,506	12,351	652	1,428
Number of manned flights	6	10	11	3	1	25
Crew members	1	2	3	3	3	Varies 2-8

Cumulative man-hours in space: 40,151 hours 54 minutes 41 seconds

INDEX